LUNAR RETURNS

TO WRITE TO THE AUTHOR

If you wish to contact the author or would like more information about this book, please write to the author in care of Llewellyn Worldwide and we will forward your request. Both the author and publisher appreciate hearing from you and learning of your enjoyment of this book and how it has helped you. Llewellyn Worldwide cannot guarantee that every letter written to the author can be answered, but all will be forwarded. Please write to:

John Townley
℅ Llewellyn Worldwide
2143 Wooddale Drive
Woodbury, MN 55125-2989

Please enclose a self-addressed stamped envelope for reply,
or $1.00 to cover costs. If outside U.S.A., enclose
international postal reply coupon.

Many of Llewellyn's authors have websites with additional information and resources. For more information, please visit our website at
http://www.llewellyn.com

LUNAR
RETURNS

●🌑🌘🌗🌖○🌕🌔🌓🌒🌑●

JOHN TOWNLEY

Llewellyn Publications
2143 Wooddale Drive
Woodbury, MN 55125-2989

First Edition

Cover art © 2003 by Brand X Images
Cover design by Lisa Novak

All charts used in this book were generated using Win*Star © Matrix Software.

Library of Congress Cataloging-in-Publication Data

Townley, John, 1945–
    Lunar returns / John Townley.—1st ed.
        p. cm.
    ISBN 0-7387-0302-8
        1. Astrology. 2. Moon—Miscellanea. 3. Human beings—Effect of the moon on. I. Title.

BF1723.T69 2003
133.5'32—dc22   2003054623

Llewellyn Publications
A Division of Llewellyn Worldwide, Ltd.
2143 Wooddale Drive
Woodbury, MN 55125-2989
www.llewellyn.com

Printed in the United States of America

DEDICATION

For Susan,

Like the Moon,
Constantly inconsistent,
Consistently inconstant,
But taken altogether,
As true as they come.

CONTENTS

CHARTS

MOONBEAMS AND MOONLIGHT: YOUR LUNAR RETURN

When John asked me if I would like to assist on an introduction for his latest endeavor, *Lunar Returns,* I was delighted and somewhat bemused. John and I have always differed in how we read and view a chart, not to mention how we arrived there!

John is steady in his knowledge and has lots of research to back up his theories and premises—me, I have a gut action and reaction. Some may call the feeling parasensitive, which is not exactly scientific or easy to explain. This is something that used to drive John to distraction. However, after many years of working on our yin and yang/female and male techniques, not to mention other issues, we have learned it is in everyone's best interest, clients included, to let John proceed first and then I can join in with great ease. As so many of our clients have told us over the years: "I cannot believe the different styles you both have and I hope I have you both on tape."

To our delights, mostly mine, our "He said, She said" still works!

John, by Sun sign astrology, is the Sun, and I, by Sun sign astrology, am the Moon.

One shines on while the other reflects.

By definition the Moon always reflects the Sun!

Hard sometimes for a Moonchild to accept! Why? Because as we know, the Moon is consistently inconsistent, or is it vice versa?!

The Sun in the meantime shines on steady.

So when it comes to the Moon, which is inconsistency continually consistent, we have a dilemma, and that dilemma as human beings is to let go and go with flow.

Just as the tides come in to the shore, they go back out again to the sea.

Just as the Moon starts small like a sliver in the sky, it grows and grows as the month proceeds until it is full. It takes center stage in the sky and then repeats the pattern or cycle all over again. This is one phenomenon we can count on (so far) just like death and taxes, as the adage goes.

The luxury of being John's partner is that John *is* astrology. He is a fearless pioneer who has a steady hand in dealing with any of the widely varied methods available and knows how to implement them into practical daily applications. When John first started using Lunar Returns with me, I was amazed at how entranced I became by my monthly Lunar Return.

I used to focus on Solar Returns (Birthday Scopes), but by seeing how great a monthly Lunar Return chart works, I could hardly wait to see how my month's chart would look.

I love detective work and the Moon is certainly worthy of that research!

This book is an easy way to gain access into the realm of Lunar Returns since John is not only a great astrologer—his way of putting pen to paper is tops. Since the 1970s, his work in the field is groundbreaking, and if you sit down and glance through the pearls here, you will find a whole new vision and vista with you and the Moon.

Whether you're eighteen or eighty-plus, this book gives you a grasp on how to use the Moon in a chart. You can dare to dream a vision, and if you want to gain the knowledge of your own inner workings, you can go for it! From decade to decade, every age will find a point of view that suits its quest—monthly, yearly, and daily.

A great cookbook is fun, exciting, has pure passion, and delivers a punch! We are able to use a great cookbook as we mix and blend all our tools and ingredients with a zeal and a passion for what we put out and what we present. This book on Lunar Returns does all that, as it glorifies the Moon. Just like the weather we see through a great set of binoculars or a wonderful telescope, the magnification of the varied meanings of the Moon are shown via the astrological chart.

So dream a dream, catch a piece of a Moonbeam, and learn how each Lunar Return assists you in impacting your options!

From an Aries Moon through a Pisces Moon, all twelve signs and houses are places you will surely want to explore . . . plus more!

Happy Moonlight sailing!

Fair Winds and Starry Skies,
Susan Wishbow Townley

PART ONE

Anatomy of the Lunar Return

Chapter One
MEET THE MOON

. . . the moon

above the rolling sea swells

to a deeper need

within each creature that must feed upon the tide

daily denied

and then fulfilled

what greater hill to climb

up the next wave of need

then to recede into decline

till all comes back again

when will it find an end

to constant repetition so ingrained that

thought cannot avoid its dictates

lest the madness of abandon fly

into the face of what the eye can but observe

the vestiges of what absorbs and still surrounds

the letters that we use to comprehend

what we would like to be an end

to up and down and right and left
forward and backward till there's
no place left to go
except to roll with what we once already knew
forever new
the tolling of the tide
outside inside
the very structure of the heart
the floating vessel that returns to meet its bride
unknown
only to the very ones to whom it's not denied
until they meet . . .

There may be only one thing that geologists, oceanographers, weathermen, sailors, fishermen, firemen, policemen, astrologers, poets, doctors, nurses, and psychiatrists all agree upon.

One thing: the power of the Moon.

This power is not theoretical, metaphorical, or mystical in nature. It is a physical power that lifts whole continents at a time, raises oceans, stirs the winds—and from all that every other effect is translated into the realms of life experience. It rocks the cradle, rocks the boat, rocks the beat, it even rocks the rocks. Short of the life-giving Sun, there is no greater force on earth that humankind and all other life must bow to.

How does the Moon do it? Simple: she is big and she is there.

In concert with the Sun, the Moon daily pulls and stretches every part of the Earth, and we all stretch with it. Moon overhead—we're pulled up and lighter on our feet. Moon below—we're heavier and sinking into the ground. It's like having alternate weights and skyhooks on our belts, twice a day.

No wonder we sometimes act strangely.

SCIENCE AND THE MOON

How strangely do we act? Take a look, first, at the words of the venerable British journal *New Scientist* concerning current research into the Full Moon effect, if there indeed is one:[1]

"Over the past 20 years, researchers looking for lunar rhythms among people have found them all over the place. Calls to crisis centres, absenteeism, heart attacks and mental hospital admissions have all been linked to phases of the Moon. Rape, robbery, assault, theft, domestic violence, suicide attempts, poisonings, drunkenness and disorderly conduct also appear to become more prevalent in the two or three days around a full Moon. A study in 1995 by psychologists at Georgia State University in Atlanta found that people ate more food but drank less alcohol when the Moon was full. In another study from 1998, a trio of Italian mathematicians looked at the timing of births. They reported 'significant clustering' of deliveries in the first or second day after the full Moon. The effect was particularly strong in mothers who had already had at least one child, or who gave birth to twins or triplets.

"What's more, survey after survey has revealed an entrenched belief among health-care workers—the people who mop up after madness descends—in the power of the Moon. In the U.S., four out of five mental-health professionals and two-thirds of emergency doctors believe that human behaviour is influenced by the Moon.

"The latest piece of evidence suggests that the lunar cycle even influences our use of technology. Last year, researchers at British Telecom noticed a 29-day cycle of peaks and troughs in network traffic. 'Just out of curiosity,' says Stewart Davies of British Telecom, 'we matched the cycle against the phases of the Moon.' The cycles coincided. In the seven days before a full Moon, people spent more time talking on the phone or surfing the Internet than at other times of the month."

It is true that dozens of investigations claim to have found lunar cycles to be associated with fertility patterns, menstrual cycles, weather patterns, plant growth, economic variations, crimes, and fires, as well as mental illness, surgical bleeding crises, plus lots more.

But, as many skeptics are quick to point out, despite many surveys, definitive evidence has yet to arrive. Similar surveys often contradict each other, some have questionable methodologies, and all lack the silver bullet that science so succinctly must supply: a physical chain of effects. That is to say, the Moon may have these widely attributed influences, but just how does she do it?

1. Gail Vines, "Blame It on the Moonlight," *New Scientist* 170, no. 2296 (June 23, 2001): 36.

The most popular theory is "biological tides," which supposes that humans and animals, being mostly water, respond in the same way that the oceans do. But that theory just doesn't hold water. The effect of the Moon's gravity upon any individual is almost immeasurably small. "The acceleration due to walking would create gravitational effects of far greater magnitude than those caused by the Moon and Sun combined," says Daniel Myers of the University of Pittsburgh School of Dental Medicine in a scathing attack on this idea in *The Journal of Emergency Medicine*. After decades of dispute over the matter, it might appear that true lunar effects can never be detected by this sort of one-on-one laboratory method.

AN ENVIRONMENTAL APPROACH

So let us try a larger, more environmental approach. Suppose you are standing on the deck of a ship. As the full Moon rises, you may be impressed by her beauty, but not by her brawn—she is not, in fact, pulling your body up at all. Yet, you are steadily rising. That's because the tide is rising, and, as they say, a rising tide floats all boats, including yours. Is the Moon affecting you? You bet it is—but by proxy, through the environment all around you. And it's not just you. Millions of creature all around you—from birds, fish, crabs, and shellfish to the tiniest microbes—are undergoing a rhythmic sea change, feeding hungrily, scurrying about making the most of the redistributed surroundings that keep them alive. This is hardly a peaceful and scenic time—it's the beating heart of life's energy exchange surrounding you while you quietly rise ever higher. No small event.

But does the Moon really affect you? That may depend on where (and perhaps who) you are. Specifics, in life, are everything—even for the ocean, which is why it took hundreds of years for scientists to actually admit that the Moon causes the tides. (Galileo, for instance, said that belief was a matter of the occult—but then he also had some serious issues with the Church.) That's because if the Moon causes the tides, the tide logically ought to rise equally at every place as the Moon passes over—but it doesn't. It rises at different rates and different times of day (even at places just a few miles apart) all over the world, usually twice a day, but sometimes only once, and sometimes not at all. Very irregular, very unscientific.

IT'S ABOUT TIDES

Despite all this, we know the Moon causes the tides. How did we figure it out? It wasn't theoretical scientists who did it, but hands-on maritime explorers, oceanographers, and cartographers who put it all together by taking the environment into account. It's a complicated business, but in brief:

As the tidal bulge produced by the combined pull of Moon and Sun rolls around the world, it encounters obstacles: bays, channels, islands, reefs, deep and shallow water. It takes real time for the huge amount of water to pass through and around these. So, for instance, if you live on one side of a narrow channel it may take hours for the water to get through, so you can have a high tide happening on one side and the other side won't see it for hours. In the meantime, the channel is frantic with roaring currents trying to get through. If you want to get technical about it, it's just Boyle's Law, simple fluid dynamics.

The result is, when you have serious topography getting in the way, you get giant devouring whirlpools like those off Scotland, Norway, and New Brunswick. And in the process, the timing of actual high tides is totally skewed from what might otherwise be expected.

And there's more. In a giant bathtub like the Atlantic or Pacific Ocean, you get bounceback. As the rolling tidal bulge swells up against one continent, it is reflected and sends back an echo that rolls back all the way to the other side of the ocean. Depending upon the shoreline, there may be several reflections that can combine with the next rollaround and make for huge tides of forty feet or more (as in the Bay of Fundy in Canada). At other times, the reflections may meet up in the middle and cancel each other out entirely, creating mid-ocean areas called tidal nodes where there are only miniscule tides or no tides at all.

It's no wonder, then, that it took so long to directly link the rise and fall of local tides to the Moon. If we had been limited to theory and statistical surveys alone, we never would have figured it out.

How much more difficult is it, then, to find that same lunar link as it affects life itself? In a word, *very*—unless and until we take a wider, yet at the same time more local, view of the phenomenon.

Suppose we do—what should we look for? Where should we look?

ALL AROUND US

The answer, probably, is just to look around us at our immediate surroundings. What are we tied into (like that ship) that is affected by the Moon and thus affects us? The sea with all its life cycles is one example, but there are more. The same rolling, bulging effect that happens on the sea also happens on land—geological tides raise the very bedrock under us on a regular basis, stressing the ground we walk on, triggering earthquakes, and causing piezoelectric effects that may account for earthlight occurrences (one form of alleged UFO). Similarly, there are atmospheric tides that gently raise and lower the barometric pressure daily and have been linked to thunderstorm activity and general rain patterns. The earth's magnetosphere also rolls with the lunar punches, varying the magnetic index, the solar wind, the cosmic ray index, and more. The fact is we live in a matrix of environmental variation affected by the Moon both on a daily and a monthly basis. We're surrounded, all-encompassed.

We're also surrounded by less "natural" forces that tend to magnify lunar effects or minimize them. When, for instance, we live in crowded conditions, the slightest variation of mood level tends to spread infectiously within the social context. This can run the gamut from a general increase in crankiness due to being around irritated and/or irritating people to actual mass hysteria when things get really out of control. Emotions are contagious and tend to compound themselves in groups. If you live in close quarters, chances are many of your natural environmental inputs, however slight, will get magnified and blown out of proportion. If you live out in the woods or anywhere expansive, however, you're on your own. You may have troubles or you may have joys, but they won't be amplified by the troubles or joys of others all around you.

We are surrounded by larger macro systems (the atmosphere, the ground, the oceans) that downshift lunar rhythms into our own middle-sized human world, which then may or may not amplify their effects. But we are also surrounded by a world of micro systems that appear to be affected by lunar rhythms as well. Experiments with various metal salts, colloidal silver, and other denizens of the atomic world have shown lunar rhythms, as have various patterns of bacterial behavior. The Moon seems to be catching us from below as well as above, from inside as well as from outside. She suffuses the very structure of our lives. Whether or not we admit to responding directly to lunar rhythms ourselves, everything else that surrounds us and even makes up our chemistry is feeling the effects—and no doubt is affecting us in the process. When everything is moving together in a matrix, so are we. Like it or not, understand it or not, admit it or not, we are carried along willy-nilly, or on a bad day, helter-skelter.

So, meet the Moon. She's everywhere, within you and without you. We dance to her rhythms, sometimes knowing it, sometimes not without knowing it, sometimes denying it, but more increasingly these days, accepting it. How much the Moon affects you specifically, however, is a matter of speculation. There may be ways to escape some of her effects—especially magnification by ill-chosen surroundings—or ways to cancel or balance them out by being aware of their rhythms and taking measures to ride the waves instead of fighting them.

RIDING YOUR WAVES

It may be more than just riding "the" waves, the environmental waves that you share with all that is around you, such as daily high and low tides, and monthly full and new Moons (when the tides run highest). It may be equally about riding "your" waves, the personal rhythms you have gotten used to over a lifetime. At the moment you were born, you climbed on board a general set of rhythms at a very specific point—a point in the daily and monthly lunar rhythms and a point in the yearly solar rhythm. If you were born at the full Moon, then that's the start of your monthly wave, your familiar starting point from which you step off anew each month. That's your most familiar stage of sol-unar tension, and it feels like home, a time to wind things up and start anew. Celestially, that's when the Moon's monthly cycle repeats the same angular relationship to the Sun it had at your birth, your own special phase of the monthly tide cycle. It happens once every 29½ days.

That's one wave. A second wave happens every day, when the diurnal swelling or receding tidal forces return to the part of their cycle where they were the instant you were born, your own special phase of the daily tide cycle.

A third wave is when the Moon returns to the place in the sky background where it was the instant you were born, once every 27½ days. In astrological parlance, that's your Lunar Return, and a chart done for that instant each month has long been believed to encapsulate the coming month, marking a starting place that characterizes the next four weeks.

ALL THREE TELL A DIFFERENT STORY

The first wave, the return of your Sun-Moon angle, marks the time of month when the overall tension of the daily tides is similar to what you were born with. Some time that day you will also experience (as you do daily) the second wave, which is the phase of the daily tide. Together, they make that day feel especially familiar, and you may feel particularly more connected to your situation, because of its innate familiarity. A monthly chart can be done for the first, and a daily chart for the second, both of which should reveal useful information about your monthly or daily outlook. Strange to say, astrologers have barely looked at the first, and at the second not at all. They are areas that beg more investigation, as they are likely crucial elements that bring the physical presence of the Moon and planets to ground in a truly causal chain of events.

THE LUNAR RETURN: WHAT IT IS, WHAT IT DOES

The third wave, the Lunar Return, is a long-entrenched astrological tool and is based on a slightly less physical way of looking at the Moon. The transit of the Moon through the houses of your natal chart is said to put particular emphasis on each house and on each natal planet it touches. When it hits your Ascendant, for instance, it makes you physically more noticeable (since the Ascendant represents your physical presence). When it hits your Sun, it boosts your ego energy, and so on with the rest of the houses and planets. And when it hits your Moon, you experience an emotional rebirthing that presages the next 27½ days of events that will determine your monthly emotional cycle. This is the Lunar Return, which is what this book is all about.

Your Lunar Return chart for any given month is easy enough to determine—most astrological computer programs will calculate it at the click of a mouse, once you have entered your natal birth data. But then it's up to you to interpret it and unlock its potential for the coming month. To do that, you need to understand its unique astrological nature.

The Lunar Return is a unique astrological beast, an ephemeral combination of a stand-alone horoscope, an electional chart (you can change it by choosing your location), and a set of transits. Ideally, it needs to be looked at in its own right, in relation to the natal chart, in relation to its own transits, and in relationship to transits to the natal chart. Its duration is too short to make its progressions meaningful, but its transits serve as if they were progressions. Unique, indeed.

Unlike the Solar Return, which encapsulates your inner position in relation to the world for the next year, the Lunar Return deals with the reflective, reactive, emotional part of your nature and thus relates more to the way you respond to events than to how you may generate them. The best use for a Lunar Return chart is in unraveling and maximizing the opportunities that events present, rather than building a game plan for future structure-building. It is tactical in nature, not strategic.

The challenge in interpreting the Lunar Return is not to plumb its depths for a vast network of details—its life is too short to get bogged down in that—but rather to extract the relevant events and eliminate the blinding chaff, to see through the smoke and dust to the immediate terrain and its possibilities. It's kind of a monthly birthday, and the arrangement of planets it displays reflects the patterns of your coming month. Each month, this "re-birth-day" works out its potential for you and then is renewed once again 27½ days later with a new set of surprises and opportunities.

HOW DOES IT WORK?

Like your natal chart, or any other kind of horoscope, a Lunar Return is a chart of a beginning—in this case, the monthly beginning of the lunar cycle that started at your birth, which is the cycle of your response to your environment, including your emotions, feelings, interactivity, social well-being, and generally how creatively you react to the challenges and opportunities of life. A Lunar Return works on the principle that when you begin something—anything—everything that flows from it is bound up in the initial conditions under which it started. The beginning is your foundation, and you build and rest upon it until you are finished. A Lunar Return is the astrological depiction of the new beginning you make each month and what results from it until the next cycle begins.

WELL BEGUN IS HALF DONE

"The beginning is half of everything," said the ancient Greeks, and so your monthly beginning is something to be taken seriously and honored, if you want your lunar month to have a special start. Attempt to give yourself some time and space, no matter how small, to rest and meditate in the few hours surrounding the time of your Lunar Return each month. Take that time to think about what lies ahead, plan your strategies, and gather your resources so you can make the best and most of what's offered. Look over the aspects in your Lunar Return chart as well as the days and times ahead where they individually kick in. Once you've got a clear, calm picture of the challenges and openings to come, you can rise to seize the day, one moment at a time, and make the most of the month from beginning to end.

CHANGING THE PICTURE

The planetary positions in a Lunar Return are locked in at the moment the Moon returns to its natal place. Where these positions fall in relation to the local horizon, and thus the areas of your life in which they work, is entirely dependent upon where you are at the time. Thus, if you see that your Lunar Return is going to develop a picture you'd like to rearrange, that can be accomplished by placing yourself at the right spot on the globe to finetune the event. Many astrologers travel widely in order to adjust both Solar and Lunar Returns—we have done so repeatedly over the last thirty-five years with great success, so it is recommended when necessary. Moving about extensively every month is not easy for everyone, but being aware that it affects the Lunar Return is a plus if you normally travel on business and have some say as to where you can go.

PIECES OF THE PUZZLE

The Lunar Return is a large piece of the astrological picture of what happens with your life every month, but it is not the only one. Lunar transits, new and full Moons, and other factors also weigh in, so they have been included in this book so you can have as complete a picture as possible of what you have to work with. Try to remember, however, that factors whose timing is as short as a monthly cycle are more like the minute or second hands on your life's clock face. Don't forget to stand back every now and then and refresh yourself with a look at the big picture of where you are and where you're going, which are described by long-term transit and progression cycles. Then you can get back down to the day-to-day nitty-gritty that your Lunar Return offers, so every moment is enjoyed and utilized to its best advantage.

With that in mind, read on, cast your next Lunar Return chart, and launch yourself into the coming month with the wind at your back and the planets racing by your side . . .

Chapter Two

INTERPRETING THE LUNAR RETURN

THE FIVE FACTORS:
FOREGROUND, BACKGROUND, ACTIONS, AND INTERACTIONS

Once your computer has spouted out your latest Lunar Return, cast for the specific location you were when it happened, it's your job to pull meaning out of it while avoiding getting lost in irrelevant details. In short, you have to separate the wheat from the chaff. It's similar to interpreting a natal chart, except that you try to stick to the highlights and pick out the trends. In a chart that only applies for about a month, the month will be over before you can go through every technique you might apply to a natal chart.

The easiest way to trim a large challenge like that down to size is to cut it up into pieces and take each piece one at a time—then return and put them all together. Here are the five factors to look at selectively as you dissect your Lunar Return.

Factor One—Your Foundation for the Month: Sun, Moon, Ascendant

As in a natal chart, the first thing you look at is the position of the Sun, Moon, and Ascendant. Their positions and aspects form the fundamental dynamic for the month: whether it's coming on like a juggernaut, sneaking up like a cat, stumbling in like a bull in a china shop, or striding in like a hero. Since the Lunar Return Moon is the same as your natal Moon, any aspects to it are also transiting aspects to your natal horoscope, thus uniquely entwining both charts and making the lunar aspects especially important. The house position of the Sun and Moon tell you where your main action will be all month, and the Ascendant sign shows the overall physical style of events. Easy aspects of these to the rest of the chart will show a month whizzing by according to plan. Hard aspects will indicate tangles and challenges to overcome. Take a look at the analysis text that applies to your specific return to see how to make the most of each aspect to get the best results from the month.

Factor Two—Personally Speaking: Mercury, Venus, Mars

The inner planets change significantly from month to month and indicate the personal permutations your life is going through. Basically, it's the framework of how you play the game, who else is playing, and how they're playing it. Easy aspects mean you win with no contest, while hard aspects suggest a more hard-fought game that may put your life skills to the test and increase your personal growth and talents through experience. The analysis text presents both sides and gives advice on which games to play, which ones to avoid, and the best strategies—whether you want to raise, hold, or walk away.

Factor Three—The Stage You Share: Middle and Outer Planets, the Nodes

The positions of the middle and outer planets—Jupiter, Saturn, Uranus, Neptune, and Pluto, along with the Moon's North and South Nodes—change little from one Lunar Return to the next, so they add more of a backdrop, the stage set on which the monthly play is performed. They are very powerful, however, and their long-running helpful and troublesome aspects are not to be taken lightly. They can be like a rope to swing yourself across the stage or a trap door that will cause a fall if you're not watching. The analysis text on the mutual aspects of these giants to the inner planets and the Lights (Sun and Moon), and where they fall by house, shows you just where to watch for them, what mood they're in, and how to get them on your side.

Factor Four—Featured Natal Positions: Your Best Foot Forward

When selected Lunar Return positions occupy the same spaces as bodies in your natal chart, this highlights certain natal inclinations for the month. It will mean that certain elements in your fundamental character will be emphasized both internally and externally. These will be the areas that best appear to represent you for the moment and that others will seize upon as entry points to your personality. These show where you can put your best foot forward to make the most effective and gratifying steps ahead, providing your first line of challenges and opportunities to embrace the world, in accordance with which Lunar Return planets hit which of your natal positions.

Factor Five—The Daily N   s

As the Moon speeds on t   nd of your Lunar Return
clock), it moves through al   hart and the Lunar Return
chart, and hits both sets o   ok for specific events that
activate these parts of your   mes for making your best
moves. There are fifty-two   k, and they repeat every
month, but not at the same   he playing field and basic
potentials of each month d   as a different role to play
every month, while retainir   all lunar transits by conjunction only, true to the "k   short-lived chart. That's
an average of one or two imp   s quite enough to see the
rhythm of the month's dail   e aspects (sextile, trine,
square, opposition, semisex   be overloaded with less
important events and lose tl   you are looking for are
these: the entrance of the Mo   12), the entrance of the
Moon into each of the Luna   tion of the Moon with
each of your natal planets (1   with each of the Lunar
Return planets (12), and the f   nar Return houses (4).
Included as well are some less   s such as eclipses, Mer-
cury/Venus/Mars retrogrades,   hemeris or transit program to see when they occur a   gly.

PERSONA   N

One very useful addition you r   pretation of the daily
progress of the Lunar Return ai   sonal Void-of-Course
Moon, something known to so   taken into account.

The period of time every 2½   s made its last aspect
to another body in the sky befor   id-of-Course Moon
period. It can last from a few r   ow late in signs the
planets are at the time. Because tl   time doesn't really have its "feet on the
ground," it is generally considered a period unfavorable for tangible decisions that
require solid support, but an excellent time for gaining insight, relaxing, and reaching

outside of confining boundaries. Many astrological calendars include a monthly list of these times, and most astrologers use this.

If you have natal planets late in one or more signs, however, an ordinarily Void-of-Course Moon may still be making aspects to your chart, allowing you a special exception from the rule and giving you a decision-making advantage at this time. Conversely, if your planets are mostly early in signs, then for you the Moon goes Void-of-Course sooner than for the rest of the world, allowing you to check out and party early, while others are still in a more mundane frame of mind. So, every couple of days when the Moon gets late in its sign, watch when it makes its last aspect to your chart, and then see what the rest of its transit through its sign is like for you. Try it out, get a feel of it, and you may see that it gives you a special edge over those who may be watching the ordinary Void-of-Course period alone.

MOVING YOUR MOON

There is a last technique for getting the most out of your Lunar Return that may be a method of last resort, but can be helpful should your Return look a little less promising than you would like it to be, which does happen. Don't just accept it, move it! As mentioned in the last chapter, you can't change the relationships of the planets to each other, but you can adjust where they fall by house and thus selectively emphasize some planets and place others on the back shelf. Unless you've got extra time and a budget to match, this will probably be a rare treat, but it can be a great excuse to take off a couple of days and change the scene—which in turn will change the stage for the whole month!

GETTING DOWN TO BRASS TACKS

The Five Factors—plus a few other tidbits—should see you through the interpretation of any Lunar Return, from the monthly overview to the daily dynamics. By looking up the analysis text for each aspect or event, you should get a pretty good picture of what's happening and what to do about it. But don't just read the paragraphs and let it go at that—use your instinct for life and all your astrology background to intuit not only what's happening astrologically but what it means when translated into the sometimes contradictory world of real-life experience. Not everything is as it seems from a purely astrological point of view, so mix in a hefty dose of the real world, because that's where you're going to live it out.

For instance, take Martha Stewart. Please! some might say—but few women have risen to such heights of business and media success. Her whole approach is lunar—it's about seeking out feelings about the soft side of the world and turning it into hard cash. On October 18, 1999, she debuted her IPO Martha Stewart Living Omnimedia (MSO), the largest of its type ever to have happened at that time. Her initial shares sold at $18 and the next day quickly more than quadrupled in public trading before settling down to $37.50 by the closing bell. Over two years later, after the dot-com crash, the World Trade Center disaster, and a recession (not to mention the troubles with K-Mart stores, which carry lines by Martha), MSO was still trading at just about its original offering price. Now that's staying power!

Does her Lunar Return just before she went public tell any of this tale? You bet it does—it's one of the more remarkable Returns any of us is likely to have. And, despite the fact that it instantly converted her into a paper billionaire, it doesn't indicate an easygoing time or even a personal home run. Let's take a look at her natal chart and the Lunar Return cast for New York City (she was in that area, exactly where at 1:44 A.M. is uncertain, but not enough to seriously affect the angles of the chart) less than four days before the public offering.

Without going into great detail about the natal chart itself, it is wise to pick out some features that might be especially sensitive to the Lunar Return of that time, considering that potentially blockbuster events were already in the pipeline. The nature of the natal chart determines how much and what kind of effect any one Lunar Return will have. A tough customer may feast off a challenging Return that a more sheltered soul might find a really rough ride. Similarly, an older and wiser individual will know far better how to use a favoring wind than one more inexperienced.

By the time of this Lunar Return, it can be safely said that Martha Stewart had become a relatively tough customer, bouncing back from an emotionally and financially damaging divorce to rebuild her former queendom into a budding empire. Her chart features a twofold division of styles, one Moon-driven and the other Sun-driven. A string of dispositorships connect a core of the Moon, Mercury, and Jupiter with a looser set of everything else but the Sun, Pluto, and the Ascendant, which form their own tight receptive square. Pluto ahead of the Sun usually means someone who never gives up, but with the Pluto-ruled Ascendant square, there's a lot of fear of getting trampled on and trampling first to ward it off—the male side of the character, so to speak, though

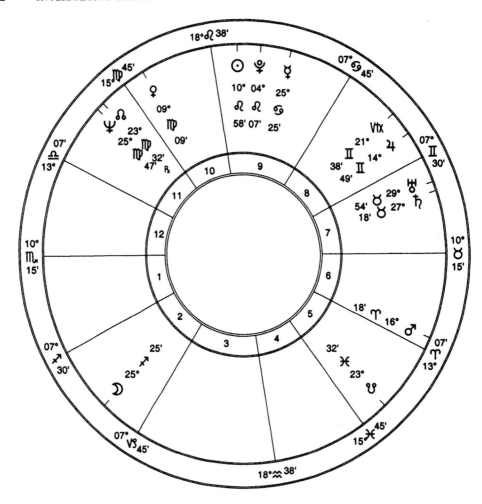

MARTHA STEWART

Natal Chart

August 3, 1941 / Jersey City, NJ / 1:33 P.M. EST

Koch Houses

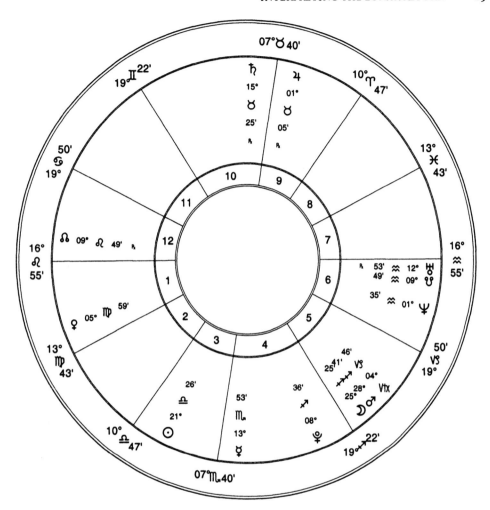

MARTHA STEWART

Lunar Return Chart
October 15, 1999 / New York, NY / 1:43:56 A.M. EST
Koch Houses

she even has the male energy to promote her career, thanks to Mars in its own sign trine the Sun and MC. The Moon in the second house, however, motivates Mercury's ideas in water (Cancer) in the ninth house, putting the creative side very much in the female camp. Moreover, the Moon is the handle along the Vertex axis of this highly driven bucket chart. Feeling manipulation of "stuff" is the focus of the lunar life, only barely connected to the solar side where the Moon barely misses a uniting grand trine with Mars and the Sun/MC in favor of a last-aspect square with Neptune. As anyone who has noted her progress on television can see, there has been a difficult disparity between creativity and control, which made her earlier appearances with other personalities an insecure disaster, something that she ultimately outgrew (no doubt a few billion dollars helped grease the path). In all regards, she is the carefully self-educated product of great talent and hard knocks, with tales of both yet to unfold.

But enough psychoanalyzing Martha—it's the Lunar Return we're interested in. This one's a classic, and it helps to know where Martha was coming from before looking at all the remarkably dense action in the Return:

Here we have all the good and the difficult characteristics of Stewart's natal chart all visiting her at once in the Return. The very first glance reveals a combination of a fixed grand cross and an out-of-element almost (again) grand trine, which are connected by an Ascendant within less than two degrees of her natal MC and a last-aspect Moon. With Jupiter and Saturn high aloft and their midpoint being the MC, it's easy to see this is wild but familiar territory for this lunar tycoon, particularly since it's all over her natal chart as well. Clearly, this is going to be one of the most challenging and rewarding months of her entire life, if she plays her cards right.

But let's slow down a bit. That's the way you look at Lunar Returns after you've done them a lot—you pick out all the major patterns and degrees that jump out at you at first glance and then fill in the details. When you're learning, you build it from the ground up as described earlier in this chapter, so let's try it that way—straight out of the analysis paragraphs—and see how it works. For clarity's sake, we'll keep the orbs fairly tight.

FACTOR ONE—SUN, MOON, ASCENDANT

Ascendant in Leo

This month is ready to befriend everybody, a real hale-fellow-well-met. It will offer a good time first, and look into things later, but it is the later that you have to watch out for. Don't forget to bring a pair of hip boots, considering what you may have to wade through, but also bring your party gear, as any occasion can be an excuse for a toast. Life is bigger than life; it's damn the details, full speed ahead. It's the little things that can trip you up, however, so don't let your enthusiasm overwhelm your good sense in the middle of the fray. When you see behind the smiles and handshakes, you can then get down to the business at hand and make the most of the moment, using the expansive energies to help roll you along. The best path: progress through joy.

Mercury Square Ascendant

Look out for crossed wires and mixed messages for a while, as what you mean to say and what you appear to be saying may be confused or misconstrued. Be ready to spend some time clarifying your statements. In this case, forewarned can be forearmed, as you'll do much better to refrain from hastily commenting on anything you're not sure of. An empty mouth beats one with a foot in it. This may involve conflicts between home and work issues, between what serves you and what serves others. You may find you cannot serve two masters but have to pick which one is truly long-term. Serve both and you may keep neither.

Saturn Square Ascendant

Be careful that you are not tripped up by a lack of self-confidence this month, so do what you know well and skip the live experiments. Repeated attempts to improve the situation will probably fail, so if you don't get it right the first time, don't beat it to death, fix it later. You may feel slighted in an incident at home or because of something someone says about you, but don't give it much weight, as you're probably working off wrong information that will be cleared up later. If you don't get the attention you want, be patient and don't take it personally. Similarly, avoid unduly harsh criticism of others right now as you will only end up with egg on your face if you turn out to be wrong.

Uranus Opposite Ascendant

Look to others, especially a close partner, for inspiration this month, as they may be a font of new and original approaches to life and circumstances. You may find them somewhat abrupt in explaining themselves, but that's impatience born of the fact that they understand what they're telling you, but you don't. Avoid confrontation while getting the details of what they have to say, and don't take it personally—it's only intended to illuminate and help. You should not only forgive personal eccentricities, but examine them and see if there isn't something in it for you, too. What may seem a strange or unorthodox approach may be just the thing needed to yield pleasure and profit, once you really get into it.

Sun in the Third House

Connections, both personal and technical, will be at the center of your efforts for a while, affording you the opportunity to be a great communicator. Now is the time to network, put your surrounding ducks in a row, and tie up personal loose ends in general. As you do this in your personal world, so you will have to do it in more common matters, particularly relating to how you are tied to the world around you by lines of communication, technology, and science. These days, you're only as good as the machines around you, so make sure everything is in working order and good repair. A flat tire or blown fuse can bring the largest operation to a halt, so go over your safety checks and make sure all is in order.

Moon in the Fifth House

A creative, even playful, response to your environment will be key no matter what your overall game plan may be this lunar month. Take time out to please yourself and be pleasing, and you'll glean fuel from it to power the rest of what you're doing. When you don't take things too seriously, that's the time when you get the most serious things accomplished, and with ease, so feel free to let go. Whatever you do, put some fun in it and it will go farther. The essence of being a successful adult is having been a successful child. Both adult and child are still available to you, so take advantage. You can't win if you don't play.

Moon Conjunct Mars

It is easy to let strong feelings reign and anger lead you into the fray—you'll regret it, so don't. That's easy to say, but hard to follow, and it involves more than just reining in your emotions—it means finding a way to channel them so you can express them and expend the energy behind them in a positive, not negative, fashion. This energy is floating about and can come back to harm you right on the spot, road rage being a shining example. Fortunately, this tends to pass quickly and not interfere in your overall operations unless you indulge in it. A rule of thumb: Count to ten, and if the feeling still doesn't pass, flee. Expend this passionate energy in love, not war.

Moon Trine Jupiter

Generosity falls on fertile ground, and giving comes easy when your heart has plenty to give. It's the feeling that counts, and the more inclusive you can be, the greater your life feast. Many of the best things in life are free right now, but you need to sit up and take notice, as it's too easy to enjoy them halfway and then let them pass. When you've got the whole enchilada, slap on the sauce and serve it up for all to enjoy. What you seize today will be tomorrow's treasure. Not only this month, but this whole year has the potential to be a pot of gold, not necessarily for the bank account, but surely for the soul. Your bowl is fuller than you thought, so dig in.

That's the foundation for the month—Ascendant, Sun, and Moon. What Martha was facing was a severe test of sheer personality—the ability to pull off in a mostly male world what no woman had done before. She had to appear to relax and shine while at the same time stiff-arming the multiple competing interests (Time-Warner among them) that would have loved to derail her debut. Had she been less seasoned, that's just what might have happened, but it didn't. That's why you always have to refer back to the nature of the natal chart. So, next we move to the personal planets, remembering we've already had a glimpse of them, as they aspected the Ascendant and Mars in this case.

FACTOR TWO—MERCURY, VENUS, MARS

Mercury in the Fourth House

Thoughts turn to home for a while, and putting things in order inside and outside your abode will be both necessary and pleasing. Rearrange things, redecorate, put a fresh look on things, and you'll rejuvenate yourself. On a deeper level, you'll be able to delineate your personal bottom line: where you feel comfortable, when you want to be alone, the foundations of your inner security. If your home is your castle, think about where it begins and the rest of the world ends, the boundaries you choose to defend and feel safe within. If home is where the heart is, define the spaces of your heart and whom you want to let in. Put up signs and indications so others will know and not transgress.

Mercury Opposite Saturn

Expect delays in even the best-laid plans and leave lots of extra room in your schedule to take up the slack. It can be easy to be pessimistic, but impatience will do you no good, and this, too, will pass if you take the time to work things out. Watch what you sign, as you could be in for more than you think and extra burdens are easy to come by right now. Don't be surprised if the cat's got your tongue and it's not always easy to express exactly what you mean. Take time to rephrase, set the record straight, and then proceed as usual. Delays are worth it only if you get it right in the end.

Mercury Square Uranus

It's easier than usual to have a sharp or sarcastic tongue, and the opportunities for cutting up with a rapier wit abound. You may find that this can cause more trouble and hurt than you think, so it's a good thing if you hold back a bit, for two reasons: First, you may be putting your foot in your mouth, as mental processes are at cross-purposes now. Second, you may create enemies who would have gladly been your friends had you been more gentle with their mistakes. Impatience leads to shallowness, however surface-bright, so follow your feelings a step deeper before letting them out of the box.

Mercury Opposite Midheaven

What you observe and learn on the home front may either interfere with or take precedence over your professional and outside life, and that may be just as well. When you forget where you came from, your external efforts cannot be sustained for long. When you renew your foundations, patch the cracks, and repour the cement, then you can sally forth with confidence, knowing your roots are solid, your support is firm, and you have refuge when you need it. Speak clearly to these issues now, and they won't pop up again later when you're too busy to tend to them. Know your heart, know your family, know your faith—then learn the rest.

Venus in the First House

You're likely to be looking especially good and will put your best foot forward. It's a great time for cosmetic improvements of all sorts to spiff up your in-person image, even though you may not think you need it. Appearance changes will likely turn out for the best, so now's the time to try out that new look. You may find that you can get by on sheer charm, so relax and enjoy it. Don't look a gift horse in the mouth. A general inclination towards affection and a loving attitude can make you the first to be chosen—all you have to do is show up and act naturally. Make sure you have substance to back it up, however—beauty is fleeting, truth abides.

Venus Trine Mars

(This one is fairly wide by orb, but completes that out-of-element grand trine.)

It will be easier than usual to let it be known what you want and have it returned in kind with love and even passion. Desire and opportunity complement each other, and whatever pleases is more than okay by everyone. You make it seem easy, even when at other times it is not, and that's part of pulling it off. The look and smell of success and ripe experience attracts more of the same, even if you've had none of either. Look and act good, and you are good. Let 'em know it was your first time some other time. Shoot first, ask questions later, and everyone will be the happier. It doesn't always work that way, but right now it does. Enjoy the bird in the hand; its heart is beating.

Venus Trine Jupiter

Whatever bounty you may have, this is the time to revel in it. The harvest of life is increased and appreciated right now as rewards are enjoyed and plans made to further increase and plenty in the future. Take more than a few moments to relax in good company, rejoice with the ones you love, and save dieting and belt-tightening for another day. There is an art to good living, so let your artistry show and share the benefits. The essence, however, is not physical wealth and possessions, but the ability to live life to its fullest, neither requiring nor asking for more than you have. When you can do that, five loaves and two fishes feed the multitudes, and the local natives show up with dessert.

Venus Square Pluto

The temptation to pull out all the stops to get what you want runs strong, and it may seem that is the only way to go, but it isn't. In fact, the more you push, the more it eludes your grasp and in the process you may destroy the very thing (or relationship) you seek to gain. On the other hand, relationships that beckon powerfully may thrill for a time and overfill with intensity, but beware when the initial glow fades—there may be little there. Passion fueled by conflict and desperate desire is often the hottest, but fades suddenly and often regretfully. Friendship fueled by patience and love lasts a lifetime. You may be called upon soon to tell the difference.

Venus Trine Midheaven

You could probably just lay back and rest on your laurels this month, but that would be a waste, since you're in position to consolidate your gains and earn yourself even more respect. You may find that self-promotion is quite unnecessary and even inappropriate, as others will do the job for you. Your function should be to appear to be as good as anything said about you, yet hardly to have worked at it. Appear to take success for granted (like it were owed to you), but in fact treasure it and take all you can get without appearing greedy. When it's your turn to shine, there's no sense in hiding your light under a bushel—let your light shine.

Mars in the Fifth House

Work hard, play hard, as they say. Well, it's time to take some time to play and put all your energy into it. You'll find it triggers that creative urge and spurs you on to greater heights. Impulsive romance is a real possibility, but take sensible precautions so you don't go further than you mean to. Keep a watchful eye on your kids, who may be particularly rambunctious and therefore get themselves in the soup. Be ready to rescue. Throughout, however, you feel good when you let yourself go and don't inhibit your spontaneity. This is a great time for sports, Xtreme vacation pastimes, or anything where having fun requires getting up and throwing yourself into it. Just remember your limits and don't try to be a superhero, even though you may feel like one.

Mars Trine Jupiter

This may be a good time to collect some of the interest on the work you have done over the years and roll it over into new investments or original projects. Anything you start now will have a good, forward-looking foundation and will easily remain abreast of the times and not become outdated. It's quite as easy, however, to just sit around and enjoy what you've got, but that would be a shame, considering what more you can do right now. It doesn't take much effort to get your increased assets working for you—that way, they'll stick around and not get frittered away. That goes for the personal side as well—be generous when you can afford to be. A helping hand is the best investment there is.

This actually paints a pretty good picture of the personal surroundings. It was an extreme time, very carefully handled, in which the entrepreneur had to cash in all her personal I.O.U.s while walking a fine line between being relaxed and charming and fending off the Wall Street wolves. Half of her act had to be warmly displaying her considerable wares and accomplishments, and the other half avoiding unnecessary confrontation while playing hardball behind the scenes with the help of her not inconsiderable allies and backers.

Next, we step back and take a look at the tenor of the times, the positions and relationships of the middle and outer planets with each other. These will tell how fertile or hostile the overall landscape may be to whatever is likely to be going on personally. In this case, we know that the economy was in the process of shooting to meteoric heights, half-real, half-illusion, and political and world changes were just over the horizon that would forever alter the future when they eventually arrived. Was this apparent in the sky? Let's see:

FACTOR THREE—MIDDLE AND OUTER PLANETS, THE NODES

Jupiter in the Ninth House

They say to keep friends, avoid discussing politics and religion—but both may come up this month with a new light shed on what you believe ought and ought not to be. Don't be afraid to discuss it, come up with new ideas and solutions, and make compromises that lead to unity. You'll feel better about yourself. You may find that your horizons expand as a result of contact with or use of things that originate unexpectedly far from home. The world is getting smaller, and your reach is getting broader. Leaps of the imagination are the order of the day, so when you run across an opportunity that seems impossible, now is the time when you may find it is really within your reach. Don't underestimate your imagination.

Jupiter Square Neptune

This is a good time for taking that second look at what life seemed to offer, or that it claimed to promise. That doesn't mean applying to your life, specifically, but you'll probably find that life's illusions are the current question in the air and there is good conversation in it at the least. It may mean for the moment that optimism fails, hopes lead to disappointment, and a lot of other confusions result which are caused by both unclear desires and fuzzy thinking. If you can be on the outside looking in on this one, you'll be the happier for it. Unexamined expectations goeth before the fall, and where you can be a cushion to others and a shoulder to lean on, you'll be the one who benefits from it later.

Jupiter Conjunct Midheaven

Jupiter high aloft makes this a great time for expanding your career position and making the world aware of how well you're doing it. Your good side is going to hit people first, and they're likely to overlook or forgive your faults, so get the most mileage of it that you can. Get your foot in the door of new establishments, even though you may not want to go there. It pays to be known. Taking an in-person approach may not be the ideal or most efficient way to do this, so use all the outside help and media tools available to you instead. That's why they're there, and that's part of this month's lesson. Spread the news, lose the blues.

Saturn in the Tenth House

This is a better time to consolidate your career position than to expand it. A careful response to criticism will go a long way to help you overcome errors and gain the trust and respect you need to make the next step. So, if people are making you the subject of the current buzz, that's a good thing. If they are, find out immediately what they're saying, as it may need to be corrected. Expect to be underestimated this lunar month and use that to surprise people when they find out how good you really are. People are only disappointed when they expect too much, so keep a low profile and then when it's time, deliver double the goods. You'll get more than double the desired effect.

Saturn Square Uranus

This is a roughly year-long period that occurs every twenty-two and a half years that, although it may not impact you specifically more than others, is usually a pretty risky time to live in. It marks conflicts across the board between radical and conservative ideologies that tend to spiral and have difficulty finding peaceful resolution. Naturally, this produces an underlying tension in everybody's life that makes harsh realities harsher and good times more desperate. If there is anything to be learned from it, it is that compromise, not confrontation, is the only way to go. Every twenty-two and a half years, an alarming number of people fail to understand that.

Saturn Conjunct Midheaven

Saturn overhead this month could get you into trouble through ill rumors and resistance in career matters. The way to avoid these is to head them off at the pass using surprise tactics. Be the first one to uncover your own errors, and you'll be said to be mature, not just mistaken. Claim responsibility for the past and then move on. When challenged, do not confront, but draw in your adversary with a quick retreat and then move on past. Intransigence will get you nowhere, so don't feel you have to prove a point. Keep your eye on what you really want in the end, not just right now. When it's clear you're in it for the long haul and have the necessary tenacity, criticism turns to admiration.

North Node in the Twelfth House

You may find this month that some of your more important commitments may be made behind the scenes and even behind the backs of everyone around you. The need for privacy about sensitive issues will probably be the reason, but it can siphon off some of the energy you would ordinarily use elsewhere. The need for you to be a listening ear and a shoulder to cry on can leave you booked, but mum's the word as far as the rest of the world is concerned until you finish helping other people resolve their personal problems and can then go back and tend to yourself. If there's such a thing as earning really good karma, this is the time for it. Oddly enough, a relatively free ride at work allows you the time to do it all.

North Node Opposite Uranus

Rocks, shocks, and potshots from left field are in the atmosphere for a bit, so keep your eyes open for something unusual coming out of thin air. This may come in the form of unusual or surprising behavior from someone you think you know, or it may be rapid developments in an unexpected area that bowl you over. Whatever it is, the ball will be in your court and you will have to return it. It might not be a bad idea to reexamine whatever seems especially normal around you, as by nature that's where surprises come from. The more open-minded and centered you are, the more likely you will benefit rather than be upset by sudden developments; to utilize their energy rather than be derailed by it.

It would seem from all this that these were promising but treacherous times during which conflicting aims were at odds for high stakes and uncertain alliances. That, of course, is exactly what they turned out to be. It was Martha's ability to play one off the other while keeping a high-profile but fiscally conservative stance that not only enabled her to go public, it also allowed her to stay steady during the next couple of years when the financial world fell apart and the sharks took to devouring each other. And that brings us to Factor Four, a return to the natal chart:

FACTOR FOUR—FEATURED NATAL POSITIONS

Any Natal Planet Conjunct Lunar Return Mercury
(in this case the natal Ascendant)

This month will tap your mental energies in such a way as to make what goes down highly interconnected with and dependent on your ability to express yourself clearly and get your ideas across effectively. It will be both a challenge and a privilege to have your point of view valued and validated by greater exposure and recognition. At the same time, expect more challenges to come from enhanced exposure, so you will need to have your arguments lined up and ready to explain when questioned or tested by naysayers. Positions you take now may well have to sustain you as the center of longer-lasting efforts, so see that you build your foundations on rock, not sand.

Any Natal Planet Conjunct Lunar Return Venus
(in this case natal Venus)

You will likely run across issues and opportunities that allow you to demonstrate your charm and magnetism and set the tone for some time to come. Expect to have both the need and the desire to captivate your audience and come away with the prize due to a combination of innate charisma and studied attraction. Getting what you want, be it money, love, position, or more altruistic gains, will be highlighted. So make sure that you not only get what you want, but want what you get, as you may have to be satisfied with it for some time to come. Be careful of what you wish for, as you just might get it, and when you do, you will want it to satisfy for the long-term.

Any Natal Planet Conjunct Lunar Return Mars
(in this case the natal Moon)

Your physical energy and get-up-and-go are put in focus this month in a way that show-cases your ability to sustain efforts and makes an impression of robust strength that engenders admiration and trust. It means, however, that you will be committing your-self to a rhythm that you will likely have to repeat, so don't design a regimen that is beyond your capabilities or will leave you drained if it becomes the expected routine. Hit hard and fast when needed, but don't use up all your ammunition on the first target you see. The essence of exercise is that it should not exhaust you, but should reinvigorate and leave you stronger for the effort. So it is with life.

Any Natal Planet Conjunct Lunar Return North Node
(in this case the natal Sun)

This month can highlight your ability and inclination to choose well and faithfully respect new commitments and responsibilities. It will probably give you some great choices to embrace, and some real clunkers to avoid. Remember that anything you do is likely to get you more embroiled than you think because it will make more waves than you expect. So don't rush into things, personal or financial, but when you do decide, do it body and soul. These opportunities will not likely go unnoticed, nor will your decisions concerning them, but don't let internal or external pressure cause you to serve up more on your plate than you can handle.

Any Natal Planet Conjunct Lunar Return Ascendant
(in this case the natal Midheaven)

How you look and your physical presence will be highlighted at critical events this month, which means if you want it to happen, show up and don't send in a substitute or a note from your teacher. Get in there and get recognized, remembering as you do that you may be establishing a new groove that you can either dance to or get stuck in down the line. So spruce up, practice your moves, and get onto the floor where you can be in the thick of things. You will be remembered for steps you execute now, at least until the next set, so step lively and step well. Put your best foot forward, and jump into the ring.

This is actually an unusual number of contacts between a Lunar Return and a natal chart. Two or three is more the average. This means that Martha was particularly drawn into the events of the month, especially as the contacts were mostly with the angles and personal planets. This is the icing on the cake that tells you things will be intense and that this may be a particularly pivotal period instead of just another four weeks chalked up on the calendar. In this case, what was emphasized and tested was Stewart's personality, which turned out to be unusually cool and steady under fire, while keeping every hair in place.

HOW ABOUT FACTOR FIVE?

That leaves us only with Factor Five, which is basically lunar transits to the natal and Lunar Return charts. This requires a rather detailed and personal examination of the month following the Return on a daily and even hourly basis. Alas, even famous people don't often leave their diaries out for public inspection, and that includes Martha Stewart. On the business side, however, new and full Moons on the Return Jupiter and natal Mars in the Return ninth house and natal sixth house perhaps describe the consolidation of company holdings and employees that resulted from reinvesting the public issue gains to buy out corporate competitor shareholders like Time-Warner.

All in all, however, just by plugging into the analysis paragraphs alone, you get over four thousand words of comment on what the lunar month should bring. After a while, you'll come to spot it all at once and you won't have to flip back and forth between sections.

As for Factor Five, in our next example we will come down chapter and verse on the development of the Lunar Return over the month, and let you flip for yourself if you want an in-depth description of every planet, house, and aspect.

The reason we can do a blow-by-blow analysis of our next Lunar Return example is that it is totally public, more public than any one of us individually could ever be, and more documented than we would ever submit to. It is the Lunar Return for New York City on September 6, 2001, which happened at lunchtime, just after quarter to one . . .

Whoa there! You mean you can do a Lunar Return for a *city*??!! Of course you can. Anything you can do a natal chart for—a person, a company, a ship, a dog, an event— you can do a Return for. Moreover, because a city's character and events are so much better documented, you can see better how well (or poorly) any astrological technique works by example, especially as we have a registered birth time for New York (as we did for Martha Stewart). No speculation here. There has certainly been no event more spectacular in the life of this most unique, soulful, and influential of cities than what happened on September 11, 2001. How does it show up in the just-previous Lunar Return? Let's take a look.

Before looking at the month's daily events, a brief look at the natal and Return charts pretty quickly tells us that something very significant is going on. Pluto rising in the Return chart is ominous all by itself, the more so because it is part of a T-square with the Sun and Saturn. Further, the Return chart has planets on both the Ascendant (Mercury) and the MC (Jupiter) of the natal horoscope, so clearly there are going to be messages spread abroad about the physical body of the city itself. Most remarkable is that the Lunar Return is only sixteen minutes of arc away from being a Mars Return as well. The city's Mars-Venus conjunction (its sparkle and charisma) is both in for a blow and a periodic renewal of energy. Clearly it's going to be an important month, though not even the most fanciful astrologer would think to suggest exactly why and how it would turn out. The chart for the time of the attack (September 11, 2001, 8:45 A.M.) is, not surprisingly, tightly linked to both the natal chart and the Lunar Return chart, but that could also hardly have been extrapolated ahead of time.

What follows here is a daily log of the astrological events hitting the natal and Lunar Return charts during the month following the Lunar Return of September 6, 2001, combined with major events affecting the city each day as reported in *The New York Times*. Each astrological event is followed by a page number to which you can flip in order to find the corresponding analysis text.

What you will find is what any given set of transits will generally reveal: a mixed bag of results. Sometimes the link between transits and events is really obvious, and sometimes you have to use your imagination. Sometimes nothing seems to fit at all—or perhaps the truly relevant events just didn't make the newspaper. You can expect just this kind of result from following your own Lunar Return through the month. You'll do best if you do not look for exact events to occur but rather for each day's happenings to fall into a given general theme.

These daily transits fall into two basic categories: lunar transits to the natal chart and lunar transits to the Lunar Return chart. Transits to the natal chart have a permanency that transits to the Return do not have. In other words, a transit to the natal chart can key off events that not only affect the month but your whole life, while the effects of transits to the Lunar Return chart are generally limited to the developing events of that one month only. For example, the Moon passing through the natal second house will have your mind on finances in general, while the Moon transiting the Return second house will put the focus on this month's budget. Sometimes in the case of the houses this will happen on the same day (when a Return chart shares the natal Ascendant), but usually not (never with the planet sets), and you get a good sense of what's important and what's merely temporary by comparing the dual sets of transits.

Another thing to take into account will be the actual critical events of the month itself, which may color everything subsequent to them. In our example, the attack on the World Trade Center is such an event, and it is the force that propelled all the other news once it took place. In the several days between the Lunar Return and the attack, you can see it's pretty much business as usual, with election issues, entertainment events, and local crime and health stories intermixed. From September 11 onward, everything was

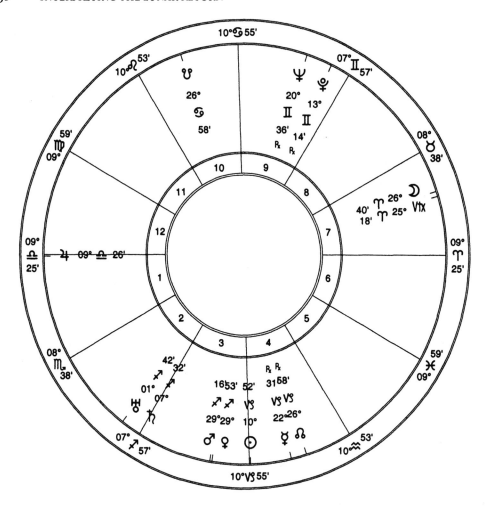

NEW YORK CITY

Natal Chart

January 1, 1898 / New York, NY / 12:00 A.M. EST

Koch Houses

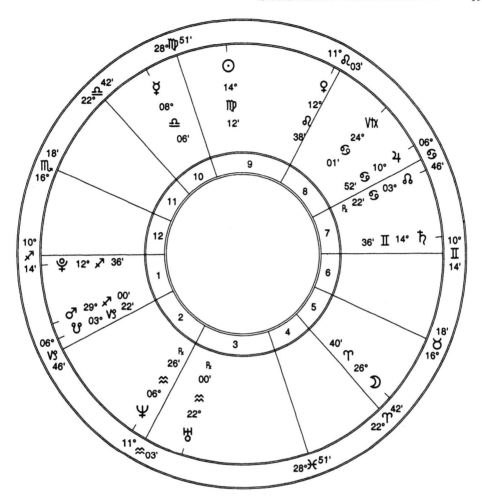

NEW YORK CITY

Lunar Return Chart
September 6, 2001 / New York, NY / 12:48:16 P.M. EST
Koch Houses

swept along by World Trade Center–related events, though still with appropriate astrological twists. By the end of the Lunar Return month, however, life was beginning to return to normal and the same topics that began the month were starting to come back into their own again. The same applies to your own life—which means be sure to take everything into context. If halfway through your Lunar Return month you get a new job, get married, have a baby, or move to a new country, any interpretation of your Lunar Return will mainly revolve around that event and be colored by it.

So, here's New York City's September 2001 saga, blow by blow, planet, and house. Events are mainly limited to things that happened in or to the city itself, even though there were many other national and international events going on, many of which were directly related to the World Trade Center attack.

September 6, 2001
Lunar Return, 12:48 P.M.

- City election campaigns are in full swing.
- Con Edison sells controversial Indian Point 2 nuclear power plant.
- The Federal Energy Regulatory Commission (FERC) limits the proposed penalties for power generators in New York who intentionally manipulate electricity.
- Mayor Rudolph W. Giuliani proposes new crime-fighting strategies.
- "Evening Stars OnStage," the series of free outdoor performances at the World Trade Center plaza, draws a packed crowd. Both focus on growing up black in the United States.
- A 49-year-old Suffolk County woman tests positive for the West Nile virus after suffering fever, stiff neck, headache, and a rash.
- Little League pitcher Danny Almonte, who became famous first for his forty-six Little League World Series strikeouts, and then for his fourteen years of age, begins eighth grade in a bilingual program in the Bronx; Rolando Paulino holds a news conference at which he admits Danny's age is fourteen and supporters continue to defend him.
- The Mets beat the Marlins.
- School Chancellor Levy says better school food will reduce truancy.

- Mayoral candidate Mark Green says his initiatives nudged the mayor into action.

- A school construction debate pits the city against the rest of state in the Albany budget.

September 7, 2001
Moon into natal eighth house, 12:08 P.M. (see page 213)

- Morgan Stanley may face a sex-discrimination suit.

- Four Democratic candidates file suit over how to count votes (to avoid a Florida-type debacle), supported by Giuliani.

September 8, 2001
Moon into Lunar Return sixth house, 2:58 A.M. (see page 216)

- *New York Times* headline: "Cruise Has Time to Repent and Time to Party." About 250 Orthodox Jews celebrate *selihot* (daily prayers approaching the Jewish High Holy Days) aboard a charter boat in New York Harbor, organized by the Jewish Cultural Fund for Performing Arts.

September 9, 2001
Moon into natal ninth house, 7:31 P.M. (see page 213)
Moon into Lunar Return seventh house, 11:52 P.M. (see page 216)

- *New York Times* headline: "If You're Thinking of Living in the Financial District; in Wall Street's Canyons, Cliff Dwellers." A profile of Manhattan's financial district as a place to live, including photos, a map, and a gazetteer of geographic data.

- Mayoral campaign: Candidates find consensus on the need for more places to live, but have different ideas for getting them.

- An abandoned newborn is found in Central Park.

- Historic temple Central Synagogue reopens three years after a fire almost destroyed it.

September 10, 2001
Moon conjunct natal Pluto, 5:08 A.M. (see page 221)
Moon conjunct natal Neptune, 6:28 P.M. (see page 221)
Moon conjunct Lunar Return Saturn, 7:37 A.M. (see page 224)

- Traced on the Internet, a teacher is charged with the 1971 hijacking of a jetliner from Ontario to Cuba—Patrick Dolan Critton, 54, living in Mt. Vernon.

September 11, 2001
Moon conjunct Lunar Return North Node, 5:04 P.M. (see page 225)
Moon into Lunar Return eighth house, 11:14 P.M. (see page 217)

- Democratic primary elections: Polls open at 6:00 A.M. and are supposed to run until 9:00 P.M. that night, but close around 10:30 A.M. statewide, profoundly changing the outcome of all the elections.
- The twin towers are attacked.

September 12, 2001
Moon into natal tenth house, 6:08 A.M. (see page 213)
Moon conjunct Lunar Return Jupiter, 6:04 A.M. (see page 224)

- New York Mayor Rudolph Giuliani warns that the death toll will be in the thousands.
- Bush labels the attacks "acts of war" and asks Congress to devote $20 billion to help rebuild and recover.

September 13, 2001
Moon conjunct natal South Node, 9:15 A.M. (see page 222)

- The National Football League calls off all weekend games, as do major college football conferences, and major league baseball postpones all games through Sunday.
- Interrupted New York primary is rescheduled for September 25.
- Some of Wall Street's biggest names search for and make deals for new space; real estate executives say some are signing long-term leases in places like Jersey City, raising the question of whether companies that leave will ever come back to downtown Manhattan.

September 14, 2001
Moon into natal eleventh house, 8:03 A.M. (see page 214)
Moon into Lunar Return ninth house, 8:35 A.M. (see page 217)
Moon conjunct Lunar Return Venus, 10:54 A.M. (see page 223)

- Bush declares a national emergency and gives the military the authority to call 50,000 reservists to active duty. The president leads the nation in prayer at National Cathedral.
- Congress approves $40 billion in emergency aid to help victims and hunt down culprits. It also gives consent for the president to use force against those responsible for the attacks.
- Mayor Giuliani declares: "The skyline will be made whole again."
- President Bush meets mud-streaked rescue workers in New York and inspects the smoking mountain of rubble where the World Trade Center once stood as he leads a grieving, angry nation in an emotional day of mourning.
- The subway and many other parts of the region's transportation network limp back to some kind of normalcy.
- The Giuliani administration, desperate to keep companies displaced by the World Trade Center disaster from leaving New York City for good, puts on a full-court press to find new space for them; big companies like American Express and Lehman Brothers reportedly have signed deals to move to New Jersey.
- In the wounded city, the school day returns to normal, and people and cars flood back into Greenwich Village and SoHo and streets through the city.
- It is revealed that a hoard of precious metals is buried beneath the rubble.

September 15, 2001
No aspects.

- The Nasdaq, New York, and American stock exchanges finish successful tests of their computer and communications systems, clearing the way for trading to resume on Monday, September 17.
- Firefighter funerals begin.
- The eastern half of the financial district reopens, but few come.

September 16, 2001
Moon into natal twelfth house, 6:26 A.M. (see page 214)
Moon conjunct Lunar Return Sun, 1:03 P.M. (see page 223)

- The New York Fire Department promotes 168 firefighters to fill the void left by the nearly 300 firefighters still missing.

- Hope fades for rescues; most may be dead.

- Cardinal Egan holds a St. Patrick's Cathedral memorial mass; 2,500 people jam the building.

September 17, 2001
Moon into Lunar Return tenth house, 12:11 P.M. (see page 217)

- Wall Street trading resumes, ending the stock market's longest shutdown since the Depression; the Dow plummets 7 percent.

- New York's Legislature convenes in a special session to pass a package of bills that would bolster the state's anti-terrorism laws.

- Major League Baseball returns to six stadiums with tributes to America and heightened security after a six-day hiatus in response to the terrorist attacks on the United States; the Mets beat the Pirates in Pittsburgh 4-1; they long for home.

September 18, 2001
Moon into natal first house, 5:01 A.M. (see page 211)
Moon conjunct natal Jupiter, 5:01 A.M. (see page 220)
Moon conjunct Lunar Return Mercury, 2:53 A.M. (see page 223)

- President Bush leads the nation in a moment of silence at 8:48 A.M. EST to mark the beginning of the attacks one week earlier.

- The Dave Letterman Show returns live from New York, a heroic tightrope walk between sorrow and stiff-upper-lip humor. (Regis Philbin and Kelly Ripa, one of New York's other main daytime talk shows, went on as scheduled throughout the crisis, however.)

- Mayor Rudolph Giuliani and other officials acknowledge that there is very little chance of finding any more survivors of the World Trade Center attacks; this grim appraisal of search efforts foretells an impending shift in operations at the site of devastation, away from careful raking of debris to a more straightforward demolition job; the "pinging" of the black box from the hijacked plane may have been detected in the rubble.

- The Yanks beat the White Sox at Chicago 11–3.

September 19, 2001
Moon into Lunar Return eleventh house, 2:31 A.M. (see page 217)

- *The New York Times:* Reports of crime fall dramatically in New York in wake of the terrorist attacks of September 11.
- New York hotels are reported nearly empty.
- Mayoral candidates debate the city's future.
- Broadway struggles with no audience.
- Stocks plummet for the second time in three days, with a late burst of buying saving the Dow from its worst three-day point loss ever.
- Giuliani's popularity soars; the president of France calls Giuliani "Rudy the Rock."

September 20, 2001
Moon into natal second house, 4:58 A.M. (see page 211)
Moon into Lunar Return twelfth house, 6:09 P.M. (applies mostly to the next day) (see page 218)

- Stocks fall sharply on fears of economic repercussions, with the Dow plunging 382 points. Bush asks Congress to give the airlines $5 billion in cash and help with lawsuits filed against them. He also appeals to Americans for their "continued participation and confidence" in the economy.
- More than a week after the World Trade Center attacks, officials announce that they will begin allowing Battery Park City residents to return to their homes; Richard Sheirer, director of the Mayor's Office of Emergency Management, says he is less optimistic about access to residential buildings closer to the World Trade Center that are within restricted areas.
- The mayoral campaign resumes.
- Manhattan criminal courts refuse a request to delay cases.
- Looting is discovered in the destroyed World Trade Center mall; it may have been done by rescue workers.

September 21, 2001
Moon conjunct natal Uranus, 9:06 P.M. (applies mostly to the next day) (see page 221)

- Wall Street stocks fall again, 14 percent. The Dow posts the biggest one-week point decline since the Great Depression (1,369.70).
- Governor George E. Pataki declares that the primary election on Tuesday will go on as planned, saying, "This is a country of laws," and that anyone wanting another term for Giuliani should "write him in."

September 22, 2001
Moon conjunct natal Saturn, 7:39 A.M. (see page 220)
Moon into natal third house, 8:29 A.M. (see page 212)
Moon into Lunar Return first house, 12:49 P.M. (see page 215)
Moon conjunct Lunar Return Pluto, 4:56 P.M. (applies mostly to the next day) (see page 225)

- In New York, residents return home to Battery Park City, near the World Trade Center ruins. The mayoral candidates resume their campaigns.
- This is the first full weekend of sporting events since the disasters, though aircraft are banned from within three miles of major sporting events and spectators are barred from taking backpacks or containers to games.
- NYC & Company, the city's official tourism marketing organization, sets up toll-free hotlines with live operators to update prospective visitors on what tourist attractions are open in the city just six days after the attacks on the World Trade Center.
- Bush tries to steady the economy "jolted" by the attacks (headline includes the Uranus keyword *jolted*).

September 23, 2001
Moon conjunct Lunar Return Mars, 11:52 P.M. (applies mostly to the next day) (see page 224)

- Giuliani may try to keep his job after his term ends in December.
- The primary candidates urge voters to turn out as an act of defiance.
- Tests find the air safe near ground zero.

- Isaac Stern dies at age eighty-one (Manhattan resident).
- Tens of thousands of people gather in New York's Yankee Stadium to pray for the missing and dead.

September 24, 2001
Moon conjunct natal Mars, 00:24 A.M. (see page 220)
Moon conjunct natal Venus, 1:36 A.M. (see page 220)
Moon conjunct natal Sun, 10:58 P.M. (applies mostly to the next day) (see page 219)
Moon into natal fourth house, 11:04 P.M. (applies mostly to the next day) (see page 212)
Moon into Lunar Return second house, 3:13 P.M. (see page 215)
Moon conjunct Lunar Return South Node, 8:19 A.M. (see page 225)

- Giuliani dismisses the possibility of a primary write-in, but leaves open other options to return.

September 25, 2001
Moon conjunct natal Mercury, 10:05 P.M. (applies mostly to the next day) (see page 219)

- Primary: Democrats Mark Green and Fernando Ferrer head for a runoff; Republican Michael Bloomberg is chosen. Dark horse Thomas Suozzi is chosen for Democratic county executive in Nassau County.

September 26, 2001
Moon conjunct natal North Node, 7:00 A.M. (see page 221)

- Stocks fall after a two-day advance, with investors wary about the economy and awaiting the U.S. response to the attacks.
- Giuliani explores an extension of his current term instead of running for a third term.

September 27, 2001
Moon into natal fifth house, 11:04 A.M. (see page 212)
Moon conjunct Lunar Return Neptune, 2:04 A.M. (see page 225)
Moon into Lunar Return third house, 11:43 A.M. (see page 215)

- The number of missing at the World Trade Center drops from 6,347 to 5,960, while the number of confirmed dead rises to 305, including 238 identified.

- Stocks are mixed on Wall Street, with the Dow and Standard & Poor's 500 index posting gains as the Nasdaq slips.

- Should Giuliani remain in office? The debate rages over the next several days.

- Green and Bloomberg say they will support a three-month term extension for the mayor, but Ferrer rejects the proposal.

September 28, 2001
Moon conjunct Lunar Return Uranus, 9:38 A.M. (see page 224)

- Officials say cleaning up the estimated 1.2 million tons of rubble at the World Trade Center site could take up to a year and cost $7 billion.

- The U.N. Security Council unanimously approves a U.S.–sponsored resolution demanding that all countries crack down on terrorism

- The Lincoln Center chief "abruptly" resigns (another Uranus keyword in headline)

- Stocks recover, up 1.9 percent; gained 7 percent for best week since 1984.

- Giuliani reconsiders a term extension after the Ferrar rejection.

September 29, 2001
Moon into natal sixth house, 10:03 P.M. (applies mostly to the next day) (see page 212)

- Officials say the entire cost of recovery from the World Trade Center attacks could reach $39 billion. That includes debris removal, overtime pay, and subway and skyscraper rebuilding.

- The New York police tally of missing at the World Trade Center drops to 5,641; the number of confirmed dead rises to 309.

September 30, 2001
No aspects.

- The New York police tally of missing at the World Trade Center drops to 5,219; the number of confirmed dead rises to 314. The death toll at the Pentagon remains 189, the Pennsylvania crash 44.

October 1, 2001
Moon into Lunar Return fourth house, 11:52 A.M. (see page 216)

- New York City Mayor Rudolph Giuliani tells the United Nations that there is no room for "neutrality" in the global fight against terrorism.
- Giuliani says he, not the state, will oversee the coordination of charity efforts.
- The mayoral extension plan loses ground in Albany.
- City officials say 5,219 people are missing at the World Trade Center, while 344 are confirmed dead and 289 dead are identified.

October 2, 2001
Moon into natal seventh house, 8:50 A.M. (see page 213)

- Faced with a U.S. economy unsettled by the attacks, the Federal Reserve cuts the key interest rate by a half-point to a level not seen since 1962.
- Authorities say 5,219 people are missing at the World Trade Center and 363 are confirmed dead.
- The city budget is battered; officials look to double aid to $40 billion.
- A salute to John Lennon is held at Radio City Music Hall.
- Environmentalists oppose a shift on Hudson dredging.
- Electrical problems cause delays on LIRR (Long Island Railroad, a major NYC commuting line).
- In general, it's back to the small stuff; business as usual.

October 3, 2001
Moon into Lunar Return fifth house, 10:56 A.M. (see page 216)
Next Lunar Return, 6:34 P.M.

- "In Little Time, Pop Culture Is Almost Back to Normal" (a story about events that occurred on October 3, 2001, that was published on the front page of *The New York Times* on October 4, 2001).

- Mayor says he won't run, but hopes for extension.
- The Democratic candidates' debate focuses on the ability to unite the city.
- A Manhattan building is evacuated after a façade bows on one side (back to a year-long issue).
- Fulton Fish Market merchants are unhappy with the new Bronx location.

Throughout the news stories, you can see themes developing as lunar transits push them along from house to house, from planet to planet. One such thread is the number of casualties, which keeps shrinking as time goes on (ultimately it was half the original estimate). Another thread is Rudy Giuliani's bid to either run again or get extended time as mayor—of course, ultimately, neither happened. You will likely find similar kinds of themes in your own life every month, as something you are especially pursuing goes through twists and turns of evolution as the Moon travels her circuit, coloring events as she goes. You will have more success putting the whole picture together if you look at the monthly transits as a rolling set of developments, not a series of discrete events. One by one, the aspects and events of the month may appear to be inconsistent or even contradictory. As Shakespeare (in *Romeo and Juliet*) says, "Swear not by the inconstant moon." At any one moment, the Moon may be inconstant, but taken altogether she is as true as they come. There is always a method in her madness, and it is up to you to spot it.

Here is a last word of advice before you launch yourself into Lunar Return interpretation: Remember, it's only a small, localized window on your life, and how it manifests is totally dependent on what's happening in the big picture. It is a way of finetuning what you're already seeing in your progressions and transits, and it never displaces or replaces them. It can, however, give you insight into just where and when the big picture will manifest itself in the details of life; where and when the general trends will get down to specifics.

PART TWO
Delineations

Chapter Three
LIGHTS AND PLANETS IN THE HOUSES OF THE LUNAR RETURN

THE SUN IN THE HOUSES

Sun in the First House

This cycle brings a strong emphasis on asserting your personality and presence in person. Life is very much live, not taped, and your successes will bloom from added self-confidence and the ability to catch your stride foremost among others. This is not an ego trip, but rather a blossoming of new internal energy that sets you apart from and ahead of those around you. Be gracious about it, but do not hesitate to take first place as the opportunity presents itself. The ability to sell yourself will be what opens doors for you, so don't be shy. It's your time to shine and bring in others to shine with you and reflect and share your light.

Sun in the Second House

The next twenty-eight days will swing around economic matters, and how you tend to them will have a great deal of influence on the way all your other efforts turn out. You will be at your best when you have laid a strong foundation that allows you to afford to do exactly what you want to do with some extra leftover for the unexpected. Do not think of this as merely a focus on the material or the mundane, but rather the opportunity to underwrite your dreams and visions so that you can afford proper attention to them as they develop. A strong foundation is fundamental to future success, so now is the time to make sure that your house is built on a rock, not on sand. Take the time to do it thoroughly and you won't ever have to look back.

Sun in the Third House

Connections, both personal and technical, will be at the center of your efforts for a while, affording you the opportunity to be a great communicator. Now is the time to network, put your surrounding ducks in a row, and tie up personal loose ends in general. As you do this in your personal world, so you will have to do it in more common matters, particularly relating to how you are tied to the world around you by lines of communication, technology, and science. These days, you're only as good as the machines around you, so make sure everything is in working order and good repair. A flat tire or blown fuse can bring the largest operation to a halt, so go over your safety checks and make sure all is in order.

Sun in the Fourth House

You may find that the next lunar month revolves around fundamental issues of what you stand for and where you live. On the outside, this may involve tending to issues of home and house to make sure that your home is indeed your castle and properly reflects and satisfies what you are and where you want to be. On a deeper level, you can find yourself looking to further uncover your roots in a deeply personal way—what you stand for, the bottom line of where your foundations lie. Issues of ethics and decisions on moral issues may present themselves, and now is the perfect time to address them, as your heart will be in it.

Sun in the Fifth House

Creativity and the more enjoyable side of life are going to be a central feature for you for a while. Let details take care of themselves, and take the time to enjoy yourself and stretch out both at work and play. When something feels right, whether it be a new idea or the opportunity to just have a good time, roll with the feeling and don't put on the brakes. Open your heart and move with the innocence of a child, and you will be rewarded for it. It is not childish to be childlike, and you have the opportunity right now to learn that it can be its own reward. Save your inhibitions for the next lunar month, or maybe longer, maybe forever.

Sun in the Sixth House

Work and health issues come into focus for the next lunar month, along with their interdependence. Observe your personal habits in order to find out how well they work for you. You may find that you have special insight into how to accomplish more and waste less time in repetitive efforts, as well as learning when to delegate tasks to others rather than taking everything on yourself. Find those small things in life that make you feel good and edit out details that are burdensome. Your daily routine has more impact on your physical and mental well-being than you might think, and now is a time when you can throw yourself into making it a support structure rather than a burden. "To labor is to pray," the saying goes. Take it seriously.

Sun in the Seventh House

It's not so much what you do, but what others can do for you that will be key this next lunar month. It's time to put your ego aside and let partners at home and work take the helm and steer for a while—you'll be surprised where you end up. Rolling with this feeling will allow you some time to tend to your inner self while the outside is handled for you. Lean on your allies and praise them for what they accomplish, as what they do credits not just them but you and your association with them. This is a time when partnerships of all kind can flourish if you throw yourself into the effort and lead with your heart.

Sun in the Eighth House

You'll find for a while that opportunity comes from resources that either you have previously forgotten or that others come in to provide. This can include attending to credit matters or finding others who are willing to back you up and throw their personal and financial weight behind you. It's also a time to rediscover lost treasures, things you had thought worthless or had mislaid. Similarly, it's also a good time to take out the trash, so to speak, and clear the decks by separating what is worth taking along and what should rightly be left by the roadside. Lighten your load, so you will be ready to take and manage opportunities and resources that come in over the transom. Make room for more.

Sun in the Ninth House

Think big and let others take care of the details: that's the central thrust this lunar month. You will have unique opportunities to observe and develop your life on a large scale, providing a bigger framework on which to hang your daily efforts. Conversely, you may find yourself involved with major external structures and belief systems, like the law and issues of fundamental ethics and morality. To a large extent, you have the ability to set your own course if you are willing to show involvement and commitment that inspires both self-confidence and trust. Be an example of what you believe and others will believe in you. The test is to be yourself and let others follow. Don't be self-conscious, just know you're right, and then go ahead.

Sun in the Tenth House

You will likely spend a good deal of time attending to what is said and thought about you this lunar month. Your reputation has great impact on how and where you succeed, so it's worth going out of your way to tend to it. It's not an ego thing, just prudent attention to publicity both professional and personal—a good time to reinforce a positive image on the world. People judge you from afar without actually meeting you, based on what you have previously done—or, just as important, on what it is said that you have done. It's a good time to sort out one from the other, so you will be able to deliver what is expected of you in the coming lunar months. Don't be shy about self-promotion, just make sure it has substance and delivers something you (and others) can bank on.

Sun in the Eleventh House

This is a lunar month to explore your support system, especially your circle of closest and dearest friends. Reinforcing loving connections will produce wonders, so don't be afraid to wear your heart on your sleeve. Let creativity flow and pass it back and forth in joint ventures, and you'll find yourself getting help from above, both literally and figuratively. Your success at this will be measured by your ability to be selfless and open and allow others to share it. Do unto others as you would have them do unto you, and you will be paid back tenfold. Seriously, it really works. The key to your own joy and passion is the joy and passion you put out. "And in the end, the love you take is equal to the love you make . . ." More, actually.

Sun in the Twelfth House

Getting beneath the surface will be the key issue this cycle, so don't be afraid to be a sleuth. What you uncover about your inner self as well as what may be going on behind your back can make the difference between success or setbacks. Cultivate the ability to recognize and understand the more troublesome aspects of yourself and those you deal with, and you will come to better know the difference between constructive criticism and intentional injury. Keep your antennae up and don't be afraid to act on what you pick up, but only after due thought about its consequences. This is a time to move with care, steady your footsteps, and look for potholes in the road of life and repair them. Your best efforts will be made out of the limelight.

THE MOON IN THE HOUSES

Moon in the First House

You'll likely lead with your emotions this lunar month, so don't be shy about letting your feelings show. Letting it all out will do you a world of good and will help delineate your personality to others who may have been working on misconceptions of where you're at. It is a strange function of letting go that you, in the end, take charge better. Holding in inhibits your ability to assert yourself; letting go gives you a free hand to use all of your personality to chart your course and get where you want to go. You will find yourself quick to react, which will give you an edge and get you in the door while others are still deciding if it's really open. Follow your intuition swiftly and with confidence, and then follow through. Being first out of the gate puts you halfway to the finish line.

Moon in the Second House

Whatever your head says about managing your money this lunar month, try following your intuition. You may get much more out of your first reaction on what to spend than a lengthy analysis and justification. Things you acquire now, however ephemeral or seemingly irrelevant, can pay off in unexpected dividends in the future. Pick up that extra duplicate before it's not available anymore, and so forth. Above all, however, remember that if you don't like what you've got, get rid of it and keep what feels best to you. You can only successfully mine the resources that feel right. Possessions are only worthwhile if they truly serve your purpose. Now is a time when you are particularly gifted at knowing what will be a support and what will be a burden.

Moon in the Third House

You may find your feelings to be somewhat scattered this lunar month, as multiple issues catch your eye and pull you in to tend to them. Don't feel you have to be rooted to any one channel of operation, but play the field so you can get a good sense of what's going on. It's a good time for putting your finger on your emotions by turning them into words, or even setting them down on paper, so you can stand back and take a second look. You'll find that there are many ways to approach a challenge, but one feels more right than all the others. That's the one to go with, but keep the other arrows in your quiver. Your ability to shift with rapid developments will be a key talent, so stay loose.

Moon in the Fourth House

You may find yourself rather inclined to stay at home and gladly let the world pass you by this coming lunar month. Certainly it's a time to gather yourself and get in contact with what you really feel and want to stick with. That needn't be carried out by literally sequestering yourself (though it may), but it will certainly be characterized by holding back a bit and not giving away the store. Protecting what is yours will be your first and best reaction, but you may have to learn to do it while carrying on business as usual. You can live in, yet apart from, the world at the same time to preserve your inner integrity.

Moon in the Fifth House

A creative, even playful, response to your environment will be key no matter what your overall game plan may be this lunar month. Take time out to please yourself and be pleasing, and you'll glean fuel from it to power the rest of what you're doing. When you don't take things too seriously, that's the time when you get the most serious things accomplished, and with ease, so feel free to let go. Whatever you do, put some fun in it and it will go farther. The essence of being a successful adult is having been a successful child. Both adult and child are still available to you, so take advantage. You can't win if you don't play.

Moon in the Sixth House

Whistle while you work, and you'll find that work is something to whistle about. You'll also find that you feel better and get more done. This is a period of keying yourself emotionally to what you do and how well you feel and integrating the two so they work for and with each other. You can find special insight into how to change your routine to achieve more with less effort by turning it into something you want to do, rather than something you have to do. Similarly, you are more closely tuned in to what is good for you and keeps you in good health. This is a time to rediscover the details of how you do what you do. Then, whatever works for you, do that.

Moon in the Seventh House

You may find this lunar month that your inner feelings and reactions are coming at you out of the mouths of others. Don't look at it as strange or invasive, but rather an opportunity to see yourself as others see you. Taking emotional cues from a partner will also allow you some inner rest and relaxation and time to get back in touch with yourself by leaning on another. Shared feelings produce better mileage than inner brooding, and opening your heart to others can be the key to opening up your life and its possibilities. The ability to know is entirely dependent upon the ability to listen well. Open your ears to other people's worlds and expand your own universe. It's a free ride.

Moon in the Eighth House

Do you have a feeling something's missing in your life? Chances are, you're right, and this lunar month is the time to find it. You've been handed a special key to the inner lost and found, so root around in your emotional attic and rediscover things you've left behind that perhaps you shouldn't have. These can be physical, like old momentos of attachments gone by, or entirely intangible, like uncovering your personal and family roots to better understand what you come from. If a purchase (or repurchase) towards this end turns out to be expensive, you may have to charge it. It will pay for itself in the long run. Oppositely, think twice about what you throw out on a whim—as soon as it's gone, you'll need it!

Moon in the Ninth House

You'll find that the more you travel, the more you know, whether your journeys are ones of the body or of the mind. This coming lunar month can put you on some new courses to unexplored territory, with a variety of interesting side trips available. Don't feel you have to plan it all rationally, but go where the mood moves you and you'll find the most interesting places. You'll find yourself standing back and seeing the whole thing and not just its parts, so don't dwell too deeply on specifics. It's the journey, not the goal, so smell the flowers along the way and gather what's in bloom. You never pass the same spot twice, so enjoy each to its fullest. After all, it's your trip, isn't it?

Moon in the Tenth House

What are they saying about you? Aren't you curious? Well, you probably should be, and now's the time to intuitively put your finger on places to shore up your reputation and beef up your curriculum vitae. Remember, it's not just what people are saying, but what they are feeling, that affects your ability to succeed, so tune in carefully. In career matters, trust your intuition and be ready to move quickly to be the first with the most. Those who hesitate are lost, so don't dawdle around finding yourself. Just make the move you feel is right, and the rest will follow. After that, you can sit back and read the reviews as they come in, knowing you've done your best.

Moon in the Eleventh House

Feelings, you've got feelings . . . and that may be the inner keyword for this lunar month, as you have the opportunity to be especially close to those dear to you. It's time to reinvigorate your special circle of friends, reinstilling and building on love that lies there. Similarly, it's a good time to expand that circle to include those who may have more to give you than they receive. You may have the opportunity to benefit from what others create, so don't look a gift horse in the mouth. Gracious receiving is as much a talent as giving, and one increases the other, so let your feelings flow and delight in what comes, knowing that you deserve it. You can repay later.

Moon in the Twelfth House

The inclination to get into yourself too much is something of a distraction this lunar month, though it can be a step along the path of inner learning. You might equally tend to what others are doing that doesn't suit your goals and take some time to remove their stones from your pathway. It's a good time to think twice before jumping to conclusions, however, so stay that first reaction, count to ten, and then jump into action—or inaction, which may be just as appropriate. Stowing your opinions can be seen as stifling yourself or merely the virtue of patience. The latter is more likely right about now, so feel free to exercise it. A carefully wrought plan works better than a quick move, so take the time to develop your possibilities before springing on them. Working out of sight, in the background, works well for you.

MERCURY IN THE HOUSES

Mercury in the First House

You may find yourself with a gift of gab more than usual, and getting your ideas across can be a primary goal. It will also be a period where you learn to articulate your own image not only to others, but to yourself. Who you are, what you like, how to look good, and how to be assertive will all take on a new and more specific light that will help you get across to others just what you want done. Avoid making it an ego trip—rather, make your personal presence and contribution a gift to the situation. The clearer you make yourself, the easier you will get on with others. Make clear where you end and others begin: good fences make good neighbors.

Mercury in the Second House

This period is marked by both the inclination and the ability to get down to brass tacks about money and finances and clearly spell out what's happening. Make lists, add up totals, put your ducks in a row and you'll feel a lot more secure about where your security is coming from. Define what's yours and what's not, so you know what you have to work with. You'll find that new ideas about money and how to handle it spring to life, and new opportunities arise as a result. This is a good time for a second look at IRAs, CD accounts, mortgage rates, investments, and the efficient use of property such as a house or car. Whatever is yours, whatever you personally own, now's the time to count it up and sort it out.

Mercury in the Third House

Brainstorming is a decided option this lunar month. Intellectual sparks fly and open up networking possibilities. The ability to put your ideas into words and get them across increases, so now's the time to write those letters, make those phone calls, chat up associates, and generally become an active part of the matrix. Dealing with technology is easier, knotty problems are simpler to resolve, and fuzzy thinking clears up. Descarte's "I think, therefore I am" applies to you double time, so put on your thinking cap and stoke your gray matter. Details fit better into the whole plan, so attend to them and be sure they are in their proper places. It's a perfect time for that "Aha!" moment when the mist clears and the landscape comes into full view.

Mercury in the Fourth House

Thoughts turn to home for a while, and putting things in order inside and outside your abode will be both necessary and pleasing. Rearrange things, redecorate, put a fresh look on things, and you'll rejuvenate yourself. On a deeper level, you'll be able to delineate your personal bottom line: where you feel comfortable, when you want to be alone, the foundations of your inner security. If your home is your castle, think about where it begins and the rest of the world ends, the boundaries you choose to defend and feel safe within. If home is where the heart is, define the spaces of your heart and whom you want to let in. Put up signs and indications so others will know and not transgress.

Mercury in the Fifth House

How to have fun? It's time to count the ways. Pick your pleasures, plan that vacation trip, and organize your creative talents so you can better express them. It's a good period to separate out work from play so you don't miss out on the latter by trying to combine them. This is also a good time for communicating with your kids, who may actually understand what you're saying for a time. In general, thoughts of love abound and just how, where, and when to express them comes into clearer view. This is one time when thoughts and feelings do not get in each other's way, but rather complement each other and run that extra mile together. Think it over, play with it, and take the time to express it to the fullest.

Mercury in the Sixth House

Organizing your work schedule will be both on your plate and easier to do this lunar month. Clear out your work space, and define what you are responsible for in relation to what your associates do. Become an efficiency expert or find one to help you out. This is also a good time to get a clearer picture of what it takes to stay healthy and define a more workable personal regimen to accomplish it. Remember that both these trends go hand in hand, and when your work runs smoothly, your health will be better. You'll be more on top of things to make both work. Prioritize, and put your health and well-being at the top of the list.

Mercury in the Seventh House

Communication with your partner or finding a partner you can communicate with will be highlighted, as will be the channels of rapport through which you do it. You can do a lot to help by concentrating on listening and better understanding where your other is at. If they don't tell you, how will you know? So encourage them to express themselves, and make your relationship clear. For this reason, this is a good time to put together business deals where you are dealing one to one. You'll be able to nail down details and avoid stepping on each other's toes later. Clarity of understanding as the result of careful expression allows you to get to central issues of the heart and the structures of living with one another, at home or in the workplace.

Mercury in the Eighth House

It's time to take a second look at those credit cards and see what you've gotten yourself into. The same goes for your mortgage or any other loans you may have. It's a good time to reorganize them to get a better deal, as you'll be more likely to spot what works and what doesn't. Clear out, consolidate, and renegotiate so you wind up with more and pay less for the privilege. Flashbacks to great ideas you once had but had forgotten about will allow you to recycle and repurpose what you already have done. Take out the trash, literally and figuratively, and separate out what you really need and can use from that which is deadweight and just taking up space in your life.

Mercury in the Ninth House

Grand ideas and schemes abound, and you could reorganize your entire approach to life if you have the time. In general, it's a space for standing back and getting a clearer overall view rather than getting lost in the details. Take a good look at the forest, and let the trees tend to themselves. You may find yourself lost in thought as you see more clearly how you fit into the scheme of things and where you want to go with your life. Put these epiphanies down on paper so you can reflect on them at leisure and reinspire yourself when life gets too crowded with mundane matters. Nail down legal issues, write contracts, and take the lead in where you're going because you know the road ahead.

Mercury in the Tenth House

Brilliant ideas about how to promote yourself and better your position in life are the order of the day, as you get a better view of what others are saying about you and where you stand in relation to the rest of the world. Clever career decisions come to mind as you sort out what's paying off for you and what's not. This is a great time to overhaul your resume or CV and make the most of your accomplishments, including ones you may not have been using. You can do more than you previously thought, and now's the time to advertise it. If you run a new flag up the pole, this is the season when it's most likely to get a salute.

Mercury in the Eleventh House

How do you express your affections with those dearest to your heart? That's a prominent question, and now is a time when you may be able to put your feelings into words to further cement what's already there. The effort is likely to be mutual, so don't forget to pay attention to what's being said, as it's deeply sincere. You'll also find this a good time for tuning in to what's being said by higher-ups, who are more likely to give you a boost as they better understand what you're doing. The key to success is to let your words match your feelings, so you function as a whole person without conflicts in the way you present yourself. A clear mind, a clear heart.

Mercury in the Twelfth House

What you say may get you in trouble, so don't lead off with the first thing that comes to mind. Put your mind in gear, and then your mouth. Behind-the-scenes efforts are favored, so you may want to keep your ideas under wrap until they can get a proper hearing. Personal insight is available especially if you are willing to put some of your hidden feelings into words and bring them into daylight. A few well-chosen words can cure inner and outer disputes as you come to better understand them. Formless fears turn to dust when you root them out and show them the light of day. Run silent, run deep—then, when you surface, take a deep breath of air.

VENUS IN THE HOUSES

Venus in the First House

You're likely to be looking especially good and will put your best foot forward. It's a great time for cosmetic improvements of all sorts to spiff up your in-person image, even though you may not think you need it. Appearance changes will likely turn out for the best, so now's the time to try out that new look. You may find that you can get by on sheer charm, so relax and enjoy it. Don't look a gift horse in the mouth. A general inclination towards affection and a loving attitude can make you the first to be chosen—all you have to do is show up and act naturally. Make sure you have substance to back it up, however—beauty is fleeting, truth abides.

Venus in the Second House

You may find yourself becoming a little more affluent for a bit, even if it's only the appearance of affluence. It's a time when you can put an especially good face on your investments and thus maximize their benefits. What you have to sell looks good and fetches a better price. What you choose to keep looks more desirable and thus increases your overall worth without further acquisition. As a result, you should strike while the iron is hot and take advantage of the situation as it presents itself. Don't part with what you love—rather, deck it out—but what you sell, sell high. Throughout, you'll be feeling a special pride in ownership, which brings admiration.

Venus in the Third House

They say talk is cheap, but right now people will want to pay for yours—not necessarily in cash, but what you have to say will be especially valued as you will be putting a spin on it that just won't quit. You may find that you have a gift right now for bringing associates together in harmony and generally spreading peace and love. You're most likely to explore areas that have to do with beauty and attraction in order to better understand and become involved with them. Flattery will get you everything, just try to be sincere. Use inner as well as outer charm. You want to do more than woo the moment—you want to be remembered for it.

Venus in the Fourth House

There's no time like the present to delve into a serious bout of interior decorating, both inside and out. That can mean changing around the accoutrements of where you physically live by painting, adding accessories, and generally beautifying your home. Or, it can mean spending some time refiguring what most pleases and satisfies you when you are home alone. On this earth, you are your own best refuge, so you should be happy taking shelter with your soul. When your inner house is in order, the world outside welcomes instead of threatens. A well-wrought starting place means a solid return. This is an ideal time to seek a balance between the inner and outer realms, so they are at peace with one another.

Venus in the Fifth House

You should find time to take pleasure in pleasure, exult in creativity, and take time out for love, whether with a partner or your family. You'll find yourself particularly inclined to let down your inhibitions in favor of a good time, and now's a good time for it. A particular talent to achieve grace and proportion will lend itself for a while, so pick up the fruits as they fall from the tree. Life is a cabaret, and for a while you're in the floor show if you'll just follow your nose. Make sure to take a lot of pictures, as this may well be a time that you'll want to cherish in days to come. The operant phrase this month: Make hay while the sun shines.

Venus in the Sixth House

Good health and what it takes to achieve it will shine on you—all you have to do is let it in. This is a good time to take another look at your diet and find ways to make it more delicious while retaining balance. The same goes for exercise—make it something you really like or would ordinarily do, rather than a repetitive waste of time. The glow of health will spill over into your daily tasks, which themselves will become more pleasurable for it. Put joy into your work, and you'll get a manifold return. When effort pleases, you'll be pleased to make the effort. The only difference between work and play is how you feel about it. Now is a time to feel good.

Venus in the Seventh House

Look for lots of love and charm radiating from your partner or, lacking that, a partner arriving on the scene doing so. In general, this is a situation where you should allow yourself to be pursued and let the other make the majority of the moves. You'll both benefit from it. Take note of how you handle yourself, as you'll likely want to look back on it and repeat the exercise with the same person or another. It's a good time to strike deals in business with equals, but not necessarily with superiors or staff. It's also likely a time when you will treat others as equals that you previously did not, and that is a good thing for all concerned.

Venus in the Eighth House

If you've not yet developed a taste for antiques and collectibles, now's your chance. One of the themes this month is lovely things surfacing from the past. It's time to take a fresh look and polish up what's gathering dust from previous days and make it part of today. Everything old is new again. If you need to put it on a credit card, that's probably doable, since there's a good chance your credit line will go up. Just watch what you do with it. Desire, and the opportunity to satisfy it, run high, just don't be greedy. Lost things find their way back to you, to your pleasant surprise. Life is a constantly changing mix of new and old—one is silver and the other gold.

Venus in the Ninth House

If you try to develop a broad sense of well-being, you'll find it surprisingly easy to connect with the higher principles of life. The miraculous way the world is put together is at your beck and call—just look up. Most particularly, you'll find yourself pleased with the way you fit into the picture and the opportunities open to you that were there all along but overlooked. Open yourself to new cultures and beliefs and you'll come away changed. Major travel and exploration, whether in the inside or outside world, are particularly beneficial. Launch out into the deep, and it will bear you up.

Venus in the Tenth House

People are saying good things about you, so you better find out what. This is a particularly good time to get as much mileage out of your reputation as possible, as the buzz about you is favorable. People are seeing you through rose-colored glasses. This may not be your fifteen minutes of fame, but it could feel like it, so treat it like gravy and make the best of it. Career moves are definitely favored—just don't let it go to your head and try to push it too far. Your best sales image is not aggressive, but generous, inclusive, and compassionate. Honey, not vinegar, does the trick. The real trick, however, is to deliver on expectations when the dealing's done. Modesty, with honesty, is the best policy.

Venus in the Eleventh House

This is an excellent time to socialize with close friends and enjoy the special relationships you have forged. You are likely to be in receipt of generosity not only from friends, but from well-placed individuals who see you as worthy of note. This is a time to bask in the light of others, so you don't have to go out of the way to shine yourself. Take time to relax and enjoy it, even when you suspect you may not quite deserve it all. Creativity runs high as a function of group interaction, so make sure you stay connected and don't try to go it alone if you want maximum benefit. For the moment, at least, it takes a village.

Venus in the Twelfth House

If you want to get your way right now, the best approach is to stay in the background and off the firing line. Your greatest benefit will come from mining your interior life, rather than trying to influence the exterior lives of others. Pleasure and accomplishment come from diving within yourself and pulling up that gnarly old oyster which, in fact, contains a pearl of great price. Personal problem-solving is favored and introspection comes easier than usual. Investments are not favored, so hold on to your coins or at least be careful how you spend them. Play your cards close to your chest if you want a winning hand, and be equally willing to hold up, fold up, or run. Or, try solitaire.

MARS IN THE HOUSES

Mars in the First House

You've got a well of energy to tap right now that will allow you to push harder, be more assertive, and go the extra mile. By exercising your strength, you can find yourself in a forward position in the ranks and obtaining a position of leadership. Avoid pushiness, but don't back off unless you see it's really the right thing to do. You have the ability to take what is yours if you are willing to flex your personality muscles. Establishing or building on a good physical exercise regimen can set the stage for better health and more energy in the long-term once you've got things underway. Don't stint, it's time to sprint. The phrase that fits: You can do it, so go for the gold.

Mars in the Second House

Put some punch into your pocketbook this lunar month. By being assertive in financial affairs, you will get more, keep more, and pick up the prize before the next person snatches it. That does not mean picking up anything you see, because you should look before you leap at this time. It does mean you have the wherewithal to put behind your purchases and will find that they have a life all their own. You may be asked to put your money where your mouth is, and you can do it on the spot, which puts you ahead of the game. But, make sure you don't promise or expect more than you can manage, or you'll be the one punched in the pocket. The operant phrase this month: The early bird gets the worm.

Mars in the Third House

You may find it hard to hold your tongue, because you have something urgent and important to say. Go ahead with it, but don't prevent others from getting a word in edgewise. Your powers of persuasion are on the rise, but make sure you really know what you are saying before committing to it. Rash thoughts and deeds are right around the corner, so avoid them. You may run into a telephone storm, or create one by getting on the horn and letting everybody know what's happening. Send out the news, but avoid spamming. Also avoid doing useless tasks, because it can be easy to get caught up with them just out of habit. Target your energies wisely so they are not wasted during this time.

Mars in the Fourth House

You can find your energies turning inward for a time, sparking you to do those more energy-consuming jobs around the house you have been avoiding. Just be careful of accidents and overdoing it out of enthusiasm once you're underway. Check smoke and burglar alarms, lightning rods, and anything that strengthens the defense systems of your home/castle. Enjoy a feeling of empowerment that comes from having battened down the hatches and being prepared for anything. You are a monarch by your own fireside when all is well with your gates. In a deeper sense, you will feel you have the power of your convictions and thus the confidence to act, or to choose not to, with equal assurance. The appropriate adage: Give me a place to stand, and I can move the world.

Mars in the Fifth House

Work hard, play hard, as they say. Well, it's time to take some time to play and put all your energy into it. You'll find it triggers that creative urge and spurs you on to greater heights. Impulsive romance is a real possibility, but take sensible precautions so you don't go further than you mean to. Keep a watchful eye on your kids, who may be particularly rambunctious and therefore get themselves in the soup. Be ready to rescue. Throughout, however, you feel good when you let yourself go and don't inhibit your spontaneity. This is a great time for sports, Xtreme vacation pastimes, or anything where having fun requires getting up and throwing yourself into it. Just remember your limits and don't try to be a superhero, even though you may feel like one.

Mars in the Sixth House

Work hard, play hard, as they say. Well, it's time to take some time to throw yourself into your work, as you have added energy to cut through challenges that may have been holding you back. That may also include an inclination to dominate your associates, however, which is appropriate only when necessary to avoid personal slights. You may feel you have to push health issues like diet and exercise, which is good, but don't go overboard. Pace yourself, as sudden strains could be self-defeating. Distribute your energy wisely and don't get hung up on a single solution. You've got the impetus to do many things well, so feel free to scatter your shot to hit multiple marks. Still yourself and focus before each move and you'll hit the bull's-eye.

Mars in the Seventh House

This will definitely not be a time to fly solo, as a large portion of your energy and efforts could come from a partner in this period, and you should take advantage of it. When offered a free ride, don't insist on walking. In fact, you will likely direct your energy toward gaining and improving relationships wherever possible, and you will have the opportunity to do so. It's important to let others make their mark, however, so avoid arguments that can come from avoidable competition or confrontation. You don't have to take up the slack, so there's no reason to try to. Put your energies into constructing a relationship for the long haul, founded on the efforts you put forth now.

Mars in the Eighth House

Old issues may come back to haunt you during this period, and you will likely dispatch them with vigor. Don't be impetuous, however, as you want to keep the tried and true, even if you've moved on to new territory. Old habits die hard, so don't kill them unless you have to. Watch what you acquire, and you won't have to dispose of it. Keep a close eye on spending and take care to have your payments in on time or your credit rating could take a hit. Conversely, if you're square with that department, this is a great time to pursue loans, credit, and investments from others. Don't tax your resources when you can let others bear the load.

Mars in the Ninth House

When given the chance to learn something new, latch on to it. You've got the energy now to push yourself into new areas of exploration that you'll use as resources later, so make room for the opportunity. Your inclination will be to follow broad outlines and let details pass, which will enable you to progress in leaps and bounds. You can go back and sweat the small stuff later. Although you may be tempted to get into legal tangles, remember that they always last longer than you think and you could run out of steam before they're over. If you get the opportunity to travel this lunar month, snap it up, because it's the perfect time. Avoid the tendency to be preachy, but feel free to broadcast your message.

Mars in the Tenth House

This is a period to shine professionally and to redouble your efforts to deserve the praise you receive. You will likely be inspired to pursue new directions with vigor and be a force to be reckoned with in the marketplace. It's a time to interact with those around you and hold your position, as to be passive would be to invite unjustified, insalubrious commentary that might do damage to your reputation. In plainspeak, repudiate rumors spread by jealous competitors. All in all, this is a good time to wave your own flag—others will rush to salute it. On the negative side, all work and no play will make you dull, indeed, so resist the temptation to become a workaholic, even though it may work for you in the short run.

Mars in the Eleventh House

Turn to your friends for inspiration and support and you'll find the extra energy there to pull you through any situation. It's a good time for intimate get-togethers with dear ones to kick relationships up a notch and cement what's already long-standing. Similarly, you may find strong support from above, both moral and substantive, which will come in the form of an unexpected favor. Eschew arguments because you'll be preaching to the choir and they don't need the aggravation. Instead, lay back into the harmony and surf the waves of friendship all the way to the beach—or to the bank, as it may happen. You should choose the position to steer and motivate, and let others run the engines.

Mars in the Twelfth House

Shooting from the hip is just the thing to avoid right now, as you'll either miss or, even worse, hit the wrong target. Take the time to scope out the situation, especially challenges that may be waiting for you in hiding, and then proceed with caution, but sure of foot. Your best work will be behind the scenes, where you can make maximum headway with minimum interference. Don't be afraid to proceed as a loner and keep secrets—they travel the fastest who travel alone. There will be plenty of time to get back into the swim of things when you've finished your undercover operations. Just resist the temptation to reveal or take credit for what you've done here, or it may backfire on you.

JUPITER IN THE HOUSES

Jupiter in the First House

This period finds you leading off with a jovial mood, and whenever you get the opportunity to expand your operations, you will. You can get others to see things on the bright side and realize that the cup is half full, and getting fuller, not half empty. You could put on a little weight without realizing it, so watch it if that's an issue. You're most likely to paint the world with a broad brush, and getting bogged down in details is what you will want to avoid. Less is not more this lunar month; more is, and the more the merrier. You may have to clean up eventually, but for the moment let the chips fall where they may to stay on the main track.

Jupiter in the Second House

You will likely have the urge to spend, and if things go as expected, you'll have what's necessary to do just that. Income and acquisition in general could be a major theme, though it will be easy to go overboard, so keep track of your cash flow so that it remains in your favor. Given the choice, favor the larger item, the single big purchase, over many small ones that more easily mount up. Putting all your eggs in one basket is the natural inclination, along with quality over quantity. Go for what will remain with you and appreciate in value so this energy will not be squandered, and you'll have it to draw upon later when you may not be able to afford to be so generous with yourself.

Jupiter in the Third House

It wouldn't hurt to touch base with everyone on your list this lunar month, as the broader a base you draw upon and the more opinions you have, the better things will turn out. Chances are you'll be running about the neighborhood a lot more than usual, so see if you can consolidate what you're doing so you won't have to take so many trips. If you didn't need call waiting before, you might need it now, as the more people you call, the more people call you back. Expanding your immediate ties to those around you will generally characterize this period, and even if not everybody has something important to say, that's okay. It's the contact itself, and the thought behind it, that counts.

Jupiter in the Fourth House

What a perfect time to expand your home surroundings—not just rearranging things, but spreading out, building up, making your tent more inclusive. You'll be in special touch with the more generous part of your nature, and to the extent you are, you will be rewarded accordingly, even if not until later. Why go out, when you can entertain at home? Parties and socializing at your place are preferable to hoofing it elsewhere, and you'll even manage to organize help to tidy up afterward. This is a fine time for sharing confidences about your inner beliefs and agreeing on what your bottom line is and what you stand for. You'll be surprised how many share your views—or are ready to convert to them.

Jupiter in the Fifth House

There's no time like the present to change into your casuals, pitch the work ethic, and enjoy yourself. Planning vacations, playtime with the kids, romance time with your partner, creativity in expanding your artistic inclinations, all should be on the front burner this lunar month. Don't feel you have to nail down the details; they will take care of themselves if you launch yourself into the effort with abandon. Don't forget to bring along your camera, as you'll want to capture the smiles and happy faces to light up your life later. Of course, you could just lay back and snooze, but chances are your energies won't let you and you'll opt for the more active approach. Don't stint, when there's fun to be had.

Jupiter in the Sixth House

This is an ideal time to try out that new diet or exercise regimen you've been thinking about. The more you do right now, the better you will feel, so keeping active will be the key to feeling good. A big shift is better than a small one, so be adventurous. Similarly, you'll find things picking up at work, where there's more to be done thanks to new developments that have arrived on the scene. When at other times you might have been exhausted, now you pick up the stamina that comes from a focus on the new and the rewarding experience it brings you. Look for it—but even if you don't, it will likely find you.

Jupiter in the Seventh House

This is an ideal period to expand your relationship with your partner or find some new ones. You'll likely discover new common ground, which puts the pizazz back in everyone concerned. Kick it up a notch, try something adventurous together, and discover unnoticed ways in which you complement each other. You'll not only find out more about each other, but you'll both grow with the experience and realize just how much potential is there if you spend some time nurturing it. On a less personal level, this is also a good time to wheel and deal on the business front and pick up new contacts that will come in and assume strong supporting roles. Just be careful you retain your full autonomy.

Jupiter in the Eighth House

Expanded credit could arrive just when you need it most this lunar month—and if it doesn't, then chase after it, because chances are you'll get it. Look for bigger credit lines, better percentage rates, and beneficial transfer deals—all are available if you look for them. The inclination will be to put them to immediate use, but there you should be more cautious and keep a substantive reserve. You'll also discover that you have forgotten resources to draw on that may show up suddenly and surprise you. It could be a fortuitous check, but more likely it will be something you had all along that you had overlooked and that will expand your horizons and capabilities. Keep your eyes open and don't toss things away without a second look.

Jupiter in the Ninth House

They say to keep friends, avoid discussing politics and religion—but both may come up this month with a new light shed on what you believe ought and ought not to be. Don't be afraid to discuss it, come up with new ideas and solutions, and make compromises that lead to unity. You'll feel better about yourself. You may find that your horizons expand as a result of contact with or use of things that originate unexpectedly far from home. The world is getting smaller, and your reach is getting broader. Leaps of the imagination are the order of the day, so when you run across an opportunity that seems impossible, now is the time when you may find it is really within your reach. Don't underestimate your imagination.

Jupiter in the Tenth House

You think no one is talking about you? Think again—everybody is, only maybe you didn't notice. Find out what they're saying and take action to take advantage of it and turn it to your benefit. It's also a great time to spread the word about yourself to potential employers, professional colleagues, and anyone you'd like to think better of you. Your best side is showing now, so make sure it's well exposed. If there is any caveat, it is to make sure that you can at least halfway live up to what others expect, when push comes to shove. Advances at work are favored, as well as praise for previous efforts and a general boost to your reputation, which, however, requires your active follow-up to make it all work.

Jupiter in the Eleventh House

An expanding circle of friends can be your biggest resource during this period, so take advantage of all opportunities to hobnob with the best. Mixing with those who can be of help to you is very much in the cards if you are willing to go out there and do it. Don't be shy—you've got to try if you want to fly. It's a good time to be especially generous with your close friends, as they have been what helped you get where you are, and they are there for you when you call. Building on that intimate base will only help shore up the foundations of your success, especially when it takes allies to get the job done. No man, or woman, is an island—hand in hand is the way to proceed.

Jupiter in the Twelfth House

Sometimes things are building up to a crescendo even when is seems like nothing much is happening. This is one of those times. The action is behind the scenes, whether you're making it or taking it. Expect benefits in the near future from events that are gestating now and that you may know little about or misjudge. Thus, avoid suspicion, because the whispers may be on your side. An invigorating swim in the deeper part of your psyche wouldn't hurt, either. You have the opportunity to open some inner gates that had been closed to you before. Take this opportunity to quietly explore those paths until you have to turn outward again. You will have your hands full soon enough, and now you've got the time.

SATURN IN THE HOUSES

Saturn in the First House

You may want to pull in your wings a bit for a while, as restricting forces are in the air that can put pressure on you if you don't carefully mete out your energies. Your personal energy reserves are easily drained, making situations you might have bounced back from take a greater toll than usual. Concomitantly, you may find that more is being put on your shoulders than you would like, so look for others to take a load off. If there is any time to delegate authority, and the responsibilities that go with it, this is it. Nevertheless, you may feel obligated to handle things yourself anyway, in which case take it slow and avoid wasting time. Less is more at this time.

Saturn in the Second House

You can't always get what you want—or at least everything you want. Now is such a time, and you will have to dole out financial resources carefully for a time and stretch your dollars with maximum efficiency. That means prioritizing, a kind of economic triage where you provide for the most important needs first and let go of what's not critical until later. What you learn and bring out of this is everything, as your ability to spend and save wisely is the very basis for later success. Don't turn away, but launch yourself into it with the knowledge that you are developing something precious that will stay with you forever: which is, essentially, knowing what's important and what's not, both internally and externally, the bottom line on which all else is built.

Saturn in the Third House

Don't worry if your phone isn't ringing off the hook for a while, because it's all part of the plan. This is a period during which you will find yourself more careful about what you say or even at a loss for words until you give the situation a second look. It may mark a quiet time in general, which gives you some time to think and regroup your mental energies. Avoid a tendency to be overly critical of your acquaintances and opt for simply winnowing out people who are time-wasters and don't carry their weight. You will learn to parse your words and thoughts so that you concentrate on what is important and let the distractions fall by the wayside. Be conservative, have reservations, doubt the usual, and demand meaning. A careful mind brings a well-wrought life.

Saturn in the Fourth House

It's time to get rid of some of those trinkets that are cluttering the house and trim the lines of where you live. Simpler is better, and every now and then more austere surroundings fit the bill. The same may go for the people who gather there—this is a time when you may find yourself cutting back on visitors and savoring your home for the secure, peaceful refuge that it can be if you let it. You may well spend some time making sure your living place is secure, both physically and financially, and if that requires extra time or sacrifice, it will be well worth it in the end. You cannot put a price on peace of mind, and the ability to go home both physically and mentally is the foundation from which you launch the rest of your life. Treat it well, and it will return the favor.

Saturn in the Fifth House

Have you been indulging yourself a bit too much lately? This is a time to take a second look at what pleases you and see if it is worth the time you spend on it or is even doing you damage. Sometimes recreation itself can become a rat race and cause you as much pressure as work itself. You may find it more useful to concentrate on serious things this lunar month and save playtime for later. If you've got kids, tell them to chill. Put off that vacation until you can afford the one you really want. In short, hoard your creative energies until the situation is right; otherwise you'll waste them. Say no to that second drink, that extra dessert, or yet another game of tennis, as you reach the point of diminishing returns. It will make the good times better.

Saturn in the Sixth House

Things slacken off in the work department for a while, though that doesn't mean there's no pressure to deliver. Quite the opposite—you'll need to put out more to get what you want done. It may be aggravating in the short-term, but in the long-term you'll thank yourself for having stayed the course. Instead of worrying about a situation that can't be helped, throw yourself into the work you have so you'll be proud of it later. Further, losing yourself in the moment is a superior way to reduce stress and anxiety and the health repercussions that they bring. When you're focused on now, you don't sweat the future, while in the process of better preparing for it at the same time.

Saturn in the Seventh House

This is a fertile time for deciding just what you'll put up with from your partner(s) and what you won't. It's a time for setting personal barriers and deciding what and who to let in, which can strengthen your relationships inside and out. Good fences make good neighbors, so set limits. In the same breath, expect to receive the same. When it all hashes out, you'll find you've chosen what and who makes you the happiest and the rest stays outside. That's why it's a good time for cementing relationships by getting married, renewing vows, and the like. You'll also find that breaking up is not hard to do; it may even be an opportunity. It's time to separate the wheat from the chaff, and go with the good stuff.

Saturn in the Eighth House

If you don't pay your bills on time, you'll suffer, and this is when you should be particularly careful to pick up every stitch. If you don't need the credit, don't ask for it, because if you're turned down, it's a mark on your record. Use what's yours rather than what belongs to others, and you'll be better off. Hoard your resources and don't pitch things you think you can't use, because you'll need them immediately the moment you do. Don't expect things from the past, but rather rely on the future to supply your needs—it will become the past soon enough. This lunar month's lesson is self-reliance, and when you've got that, all the rest is gravy, so pitch in and make it so.

Saturn in the Ninth House

Grand ideas and sweeping generalizations are not of much use right now—you're better off getting back to basics. Greater demands will be made on your sense of right and wrong, so the more in touch with yourself you are, the more you can rise to the challenge. You'll be better off working from fundamentals that can't brook argument, as excuses won't be tolerated, or shouldn't be. Stay out of the courts, stay relatively close to home, and work on making what you've got better—no frills, just what applies and no more. Get your feet on the ground and keep them there for a while. Speculation and flights of the imagination can wait until you've redefined your bottom line. Omit needless words, and don't repeat yourself. If it's sufficiently simple and streamlined, you will get your point across. If not, trim it.

Saturn in the Tenth House

This is a better time to consolidate your career position than to expand it. A careful response to criticism will go a long way to help you overcome errors and gain the trust and respect you need to make the next step. So, if people are making you the subject of the current buzz, that's a good thing. If they are, find out immediately what they're saying, as it may need to be corrected. Expect to be underestimated this lunar month and use that to surprise people when they find out how good you really are. People are only disappointed when they expect too much, so keep a low profile and then when it's time, deliver double the goods. You'll get more than double the desired effect.

Saturn in the Eleventh House

Whom do you really care about and who is just a pleasant acquaintance? This might be just the time to ask that question and pare down your circle of close friends to those you most truly love and trust. And further, since absence makes the heart grow fonder, simply staying out of company for a while can give you the answer. Don't expect too much in the way of moving more upscale right now, as times are changing and you're not the only one retooling your surroundings. Pay more attention to what you've got than to what you'd like to have. You may find that the greatest treasures are already in your grasp, and this is the time to secure them and be sure they stay around.

Saturn in the Twelfth House

If you think you got away with it scot-free, take a second look. Failings you may have forgotten about can come back to haunt you right about now, so it might be a good idea to go back over the recent past and pick up any stitches you might have missed. The issues that will slow you down are sometimes the most difficult to see, so welcome the input of others, even if it sometimes may be less than flattering. You can't fix what you don't know is broken. Consolidate your position behind the scenes so you don't give yourself away by showing loose ends. If you write in a little extra time on your calendar for unexpected delays, you'll still be right on schedule.

URANUS IN THE HOUSES

Uranus in the First House

It's a time to expect the unexpected, but the lesson this lunar month is to avoid losing your temper over it. Sudden changes, reversals of field, and distracting demands might drive you crazy if you didn't know they were coming so you can keep your cool. And if you have a couple of quick changes of heart, don't worry about it; it goes with the territory. Don't expect others to be so tolerant, however, so have an explanation ready at all times. Although all this is usually mental and emotional, look both ways before you cross the street. It's easy to get lost in thought and end up facing an oncoming surprise you don't want. Nevertheless, it's exciting, because sudden breakthroughs are the heart of the spiritual experience.

Uranus in the Second House

You might want to spend some extra time balancing your checkbook this lunar month, with an emphasis on keeping a cushion to avoid surprises. Sudden expenditures can take you by the lee and then be followed by equally unheralded infusions of cash. Don't let it upset you, as it all evens out in the end, but it can be a bumpy ride until then. At the same time, it will open up new ideas of income possibilities as you discover there are more resources out there than you thought. You don't have to develop it, it will just appear out of the blue. All you have to do is stay ready and steady, pending its arrival. Of course, that may be easier said than done.

Uranus in the Third House

Flashes of insight, bolts from the blue, and startling messages on your answering machine all could be around the corner. You may experience sudden interruptions of calls and meetings, stop-and-go traffic that is heavier than usual, computer crashes, and events bunching up together with long, vacant spaces between. It's a time of spotty and ragged communication, but when you get your point across, or get the message, you can unexpectedly find yourself on new ground. Don't try to keep it under too much control, just ride with it and see where it leads you. There are no times like interesting times, and here they are, so enjoy them. Don't make dates you can't break, or you'll be forced to break the dates you make. Just free your mind and see what happens.

Uranus in the Fourth House

Try not to stray too far from home, because you may suddenly be called back there to tend to a new situation that has unexpectedly arisen. On the physical side, check your smoke alarms, burglar alarms, lightning rods, and other safety systems so the shocks you get won't be catastrophic—then sit back and see what happens. It's bound to surprise you. On the personal side, you may find yourself in receipt of sudden epiphanies concerning the very stuff you're made of, visionary insights about what provides your stability in life. These may disappear as quickly as they came, so keep a pad and pencil handy to jot down what arrives for use later. Those "Aha!" moments slip away too easily unless you tie them down.

Uranus in the Fifth House

Creativity comes in fits and starts and may interrupt your routine, but let it. Be ready to catch unexpected opportunities for enjoyment and relaxation, and don't tie yourself down so much that you can't be swept away should the opportunity present itself. If you have to drop everything, do so. You can pick up where you left off later. Don't look a gift horse in the mouth, even if you think it's not one. Treasures lie buried in the strangest of places so don't jinx it by being a stick-in-the-mud. Expect surprises from your kids (that's not unusual), but be ready for a quick rescue on a moment's notice. Most of all, don't get frazzled, or you won't have any fun. Surf it.

Uranus in the Sixth House

Just when you thought work was boring, everything changes, without notice. Maybe you'll look back with fondness on boring, maybe you won't, but chances are you won't have time to. Success depends on keeping on your toes. Don't try to stay ahead of the curve lest it vanish beneath you unannounced—better to ride the crest of the present and stay flexible. The challenge of it all will be avoiding stress and anxiety in the process, which can impact your well-being. Knowing that each up and down is only temporary helps, and it will all even out in the end, particularly if you focus on a quick response now and let later take care of itself. That's the way surprises turn to opportunities, so don't recoil—embrace them.

Uranus in the Seventh House

It could be really easy to get into an argument with your partner around now, but don't fall for the bait. Just back off, and things will subside as quickly as they arose. The same kind of electricity in the air makes for sudden spontaneity, and you can get into things together you never dreamed of before. You also may find that a new relationship drops down on you from nowhere and demands your immediate attention—which is, of course, just what you should give it. These things don't happen every day. Don't: get cross, get mad, fight back, get hurt. Do: be spontaneous, cut loose, experiment, be free. These things have small windows of opportunity but big possibilities. Don't let them evaporate before you're fully aware they're there. Stay alert, react quickly, and be ready to cast your fate to the winds.

Uranus in the Eighth House

If you were going to win the lottery, this would be the time. Or, you could equally as well find you'd forgotten to pay some important bill and it's come knocking. When you're dealing with credit or looking for new financial backing, surprise is the order of the day and it may be difficult to keep the situation in the usual order. So, don't expose yourself to unnecessary financial risk right now, but also be very alert for a windfall that could slip through your fingers if you aren't on top of the situation. On the personal level, you may expect a sudden blast from the past, which could be an old flame who is too much older or a playground bully who has become a priest. Whatever it is, it will surprise and amuse.

Uranus in the Ninth House

Lightning strikes again. If you're willing to take the time out for some introspective philosophizing, your world could be turned upside down this lunar month. Don't be surprised if your moral assumptions are suddenly challenged or you find yourself converted to positions you might never before have espoused. An open mind is the key to success right now, so be willing to consider every side of every situation without bias or predilection. Be equally ready to have your own presumptions shot down, and abandon them with grace. Sometimes it takes a long time to learn, sometimes it all comes in a flash of realization. Right now the latter rules, so make room for it. Truths you embrace now will support you later.

Uranus in the Tenth House

When you're high on a ladder, you are most likely to reach over and gain the top or trip and fall. A rather sudden dose of either or both may be in store in the career department, so buckle up your safety harness just in case. Sudden gusts of fame can shore you up or an unexpected twister whirl you around. If you give yourself some slack and back up, you'll rise. If you climb blindly, you'll fall. Let occasional harsh words spoken about you roll off your back and take sudden praise with grace. When all is said and done, your reputation will follow your deeds, so attend to work and not to ambition, and let the buffets of fortune strike your exterior while the calm remains within. Be unflappable.

Uranus in the Eleventh House

Creative insight is right around the corner if you are willing to take a cue from your nearest and dearest and tag along with their bandwagon. Avoid challenges without first thinking them over, and know that other's success happens for a reason and they will help you gain the same if you listen. Leave the kids with an uncle or aunt and spend some time in good company with discovery on your mind. The doors of perception are only next door—just walk on over and spend some time there. Be quick to pay heed, but slow to argue. When someone offers to drive, let them. Free rides right now come where you least expect them, and they'll pass right by if you're only listening to the sound of your own voice.

Uranus in the Twelfth House

This is a good time to look twice before you accept unexpected offers or suggestions to switch horses in midstream. Snap decisions can be your undoing, so slow and easy wins the race, even if the hare seems to race past you. If it looks too good to be true, chances are it is. If you can't check out where it comes from, be extra wary. The uninspected opportunity has troubled roots, so look carefully, and dig to find them. The roots may be your own, so don't get hoisted on your own petard. As Pogo said, "We have met the enemy, and he is us!" Question others, and question yourself. The answers will be revealing and will have made this period very worthwhile.

NEPTUNE IN THE HOUSES

Neptune in the First House

You may find you have some difficulty making yourself clear this lunar month, so don't be surprised if you have to repeat yourself to make a point. On the other hand, your ability to dissemble is heightened and if you need to pull the wool over someone's eyes, this will most certainly be the time. People will tend to project their own fantasies on you, and your best response is to work with what they see rather than to try to clarify things, as you may have difficulty doing so. It's a good time to get caught up in your imagination, with yourself as the main player. If you like the part, keep it. If not, choose another. Things are very fluid right now.

Neptune in the Second House

It may be hard to keep a firm grip on your finances for a time, and if you want an exact bank balance, total it up several times. Forget exact certainty in that area and just ballpark it until the month is over. It's a good time to use free association to discover new sources of income, but not yet the time to act on what you find out. Don't believe everything that's promised in that area, but expect that some of it will indeed work out well, while the rest vanishes into thin air. Speculation runs rampant, but don't bet on blue-sky investments that can wither under the light of scrutiny. Be skeptical, but don't allow that to interfere with your creativity. After all, "If you don't have a dream, then how are you going to have a dream come true?"

Neptune in the Third House

If you get a lot of wrong numbers for a while, don't let it bother you. It's just fogged-out people who keep missing the buttons, of which you may be one. It's a great time to brainstorm with your buddies, "what if" being the central theme. Just don't insist on specifics, but let your mind run free and nail down the details later, if your inventive solutions ultimately lend themselves to it. Better living through chemistry may help fuel it all, but remember to separate pipe dreams from hard plumbing in the end, for smoke rings and currents in a glass can lead you astray as well as inspire. Write down good ideas, as they will tend to escape you, and some of them could be the start of something really new and important.

Neptune in the Fourth House

If you're tempted to lie around the house and just dream sweet dreams, this is an ideal time for it. Be it ever so humble, there really is no place like home, and you'll likely find some of your most creative moments in solace, communing with yourself in comfortable and undisturbed surroundings. If you're inclined to change things around, however, think about it, outline it, formulate it, and then put it off for another time, as this is not a great time for implementation on the home front. Better to give the imagination full rein and envision how you'd really like to live, and then see if it's a practical possibility when you're in a more lucid frame of mind. First fantasy, then reality.

Neptune in the Fifth House

Your most creative moments will come when you abandon yourself to fantasy and float free, allowing anything at all to come into your head and let the Muse do her work. Breaking down internal barriers, however you choose to do it, will be the key to finding that field of dreams that had escaped you. If you build it, they will come, but this is not really the time to start construction. Just outline your inspirations and wait until your head clears before you get down to brass tacks. Your kids may try to pull the wool over your eyes, but what else is new? This is also a fruitful period for planning that dream vacation, just don't book your tickets until you're sure that's really where you want to go.

Neptune in the Sixth House

Health issues may temporarily cloud your skies, but don't let your imagination run wild—there's a concrete explanation, and solution, waiting in the wings. Don't make decisions in that division without consulting an authority, as flights of fancy may lead you astray and exacerbate the situation. Everyday work proceeds as if in a dream, which is about where you should leave it for a while, as rearranging your routine now will simply mean you have to do it all over again all too soon. It's not that you can't come up with some ideal solutions, just that they may not seem so ideal as circumstances change down the road. In general, let sleeping dogs lie for the moment and tend to thoughts of improvement that can be put into action in the near future.

Neptune in the Seventh House

If communication with a partner sometimes seems elusive, it may be because you are letting your own projected image get in the way. The temptation to make another into something he/she is not, but seems to be, is easy to succumb to. Knowing that, this is a good time to explore the illusion and come to better understand why you have it and how to correct it. On the other hand, it's also a good time to use the illusion and transport mutual feelings to a more ideal level. It's not a great time to settle on a new relationship, until you really know what's happening, but a great time for getting into one and exploring the possibilities. What you don't know can hurt you, but you've got time to winnow out fatal flaws and correct them.

Neptune in the Eighth House

Don't let your credit cards run away with you this lunar month, or they're likely to run away from you later. Prospective purchases are not all they are made out to be, so don't buy a pig in a poke—make sure you've got a warranty and can return it. Prospects of easy money, great interest rates, and fabulous bargains may abound, but read the fine print. If it looks too good to be true, chances are it is. Lose yourself in the past instead, rummage through the attic where fond memories revive, and dust off the old and figure out ways it can be made new again. If you're tempted to dispose of seemingly useless items, wait a bit. They are not as useless as they seem. What is broken can be repaired, and what is old can be made new. Imagine the possibilities.

Neptune in the Ninth House

You are only as big as your dreams, and this is a good time to formulate them. Your ability to ignore the trees and encompass the sweep of the forest is heightened, and you'll be able to distill the essence of situations that might have been too detailed to stand back and analyze. On the other hand, where standards are already laid out, you can get tripped up by technicalities by which you would ordinarily abide as a matter of course. It's probably a good period to avoid mundane entanglements and spend your time thinking about the deeper issues of life, where you can set the time aside. Visions you first sight now can become beacons for your life in times ahead, so keep your eyes, and mind, open.

Neptune in the Tenth House

You would do well to go out of your way to squelch rumors about yourself, and nip wrong ideas in the bud before they can do damage. On the other hand, because you are likely to be thought about inaccurately, it's also a good time to feed overblown myths of your greatness and accomplishment. It is not a good time for job-seeking by sending in a resume, which may be misinterpreted—far better to show up in person. If you're thinking of changing or modifying your career, it's a great space for dreaming up new things you could do well and hadn't thought about before, but not so good for throwing yourself into a new direction unless you have checked it out thoroughly in advance. Nevertheless, the operant phrase is: If you can dream it, you can do it.

Neptune in the Eleventh House

This is a wonderful period for gathering old friends together, kicking back, and partying down. Memories will be tinged with a pleasant soft focus, current events viewed through rose-colored glasses, and the future painted bright and full of prospect. Don't kid yourself that you're coming up with practical analysis or concrete solutions, but the ideas run free and the company is warm, so what more could you ask? You're best off with people you already know well, as bringing new blood into your old circle may just cause confusion. Don't try to garner support from higher-ups, as you'll only muddy your prospects. Stick to where affection already lies and can be reaffirmed.

Neptune in the Twelfth House

Dreams, visions, fantasies—all may fill your head, especially on awakening in the morning. They can give you special insight into your feelings, but don't take them literally or act on them without good reason reconsidered later. They are there to inform, not to follow. Similarly, in the outside world you should keep a watchful eye out for deceivers and those who would go behind your back to take advantage of you. Simply shining the light of your awareness of them dispels the possibility of harm, but don't be caught napping. Be selective, however, as you may mistake well-meaning dissembling for an intentional misdeed. For instance, don't spoil your surprise party, just because you got wind of it!

PLUTO IN THE HOUSES

Pluto in the First House

The keys to getting your way this lunar month likely lie in your ability to strongly project your own personality in such a way that it overpowers others. This can make you the dominant alpha player in your surrounds, but you may pay the price of resentment by others who feel pushed out of the way. A better way to approach this is to distribute extra credit to other players while quietly seeing that you have achieved your goals, and only your goals, in the subtlest possible manner. You could win in a shoving match, but it's better if competitors get at least a part of what they want, while you still retain the prize most important to you. The iron fist in the velvet glove is as obvious as you should get.

Pluto in the Second House

You may find money issues to form something of a battleground, and your use of them will give you a competitive edge. This does not mean you should use your financial clout, such as it is, as a bludgeon—rather, by supporting the cause you feel strongly about, you use your advantage for the general good. Similarly, expect power plays in this area and be ready to pay your own way if challenged. This can transform your view of how you can use your possessions, but take care to avoid short-sightedness in procuring what you want at the expense of others. Temporary triumphs can come back to haunt you, and gracious generosity will be paid back manyfold, if you are ready to compromise.

Pluto in the Third House

Words can be your tool, or weapon, of choice this cycle, so be careful how you use them. This is the playing field upon which confrontations are most likely to be staged, but also where you have the opportunity to transform conflict into compromise. The pattern this will take will be less a mutual acceptance of the status quo than a willingness to throw that out entirely and forge a new way to phrase, and thus execute, the matter at hand. The pen is truly mightier than the sword, but swords are better put to use when reforged into ploughshares. A cruel but clever wit can clear the playing field, but leave its possessor alone at the end. Don't push it over the edge unless irrevocably pushed into it by another.

Pluto in the Fourth House

If you have any simmering concerns or disagreements with others on the home front, this is not the time to bring them to a boil. If you avoid the temptation to rise to the bait, you'll find that seemingly intransigent problems simply fade away by themselves through changing circumstance, leaving you to put together a new order in harmony. Opposing a sea of troubles will not end them, so it's better to weather it out and wait for a change in the surroundings. If this means compromise, go for it, or simply refuse to get into a fracas at all. Remember, your home is your inner countryside, and civil wars do the most to ravage it. Better to tarry than be stern. All things come to those who only sit and wait.

Pluto in the Fifth House

Even the lightest of games can be taken too seriously, so make sure that the subtext to friendly competition doesn't involve ego-risking conflict. It's only a game, and you can win it this cycle, but consider at what cost. The same applies to anything you do for pleasure—love, sports, relaxing with children, vacationing—remember to let others have their way, as long as they're not stepping on you. It takes more than one to play, unless it's solitaire, and it's the game that counts, not the victory or loss. Make sure everybody comes away feeling good about themselves, and if you must force an issue, force that. Good times are the source of endless happy memories, but pleasure selfishly derived spawns a residue of regret. No regrets.

Pluto in the Sixth House

The workplace offers you a number of opportunities to do battle this lunar month, but if you are aware of it, they are easily avoided. Avoidance is the right response, as the issues at hand will die of themselves and another, better, situation will arise if you let it. The bottom line is, everyday affairs are not worth going to the mat about, even if they seem so at the time. Further, the pressure that confrontation brings has its greatest impact on your health, as stress and anxiety bollix up your system and throw beneficial habits to the wind. There are no life-and-death issues at stake, really, so don't opt to make your last stand, or it could be Custer's. When you flex your muscles, do it only for sport.

Pluto in the Seventh House

This could be the time when you find yourself drawing the line with your partner, so good-faith negotiation and fair play are especially in order. Don't be a doormat, but don't dominate either, even though some issues seem only responsive to an either/or solution. Win/lose is not the way to go; win/win is. Lacking that, simply put aside unresolvable situations and deal with them later, when you will likely find them no longer relevant and well-disposed of, and not worth getting bent out of shape about. Respecting boundaries is what it's all about, and the best way to avoid a border war is to back off from the border. For any relationship, a temporary DMZ beats a final DOA, and allows both sides peace with honor. When in doubt, call a truce.

Pluto in the Eighth House

Memories of the past may tend to overwhelm you during this lunar cycle, but you can learn much from them. Remember that what is lost is found again in time, and every day is a new chance to reshape your life. Conflicts over family property can be resolved by sharing, so no one takes the lion's share but each gets what is personally special. Learning to give up what you desire for the moment is the lesson of the month, and what you permit to pass will, in time, come back. What returns is truly yours, and what doesn't, wasn't meant to be to begin with. The opportunity to place the importance of spirit over matter is knocking, so don't bar the door.

Pluto in the Ninth House

If there is a time to pass on discussing religion and politics, this is it. Not that you can't be sufficiently eloquent to command the field, but that is likely to be to another's sacrifice. It's only a point of view, so what's the point? On the other hand, this is a good time to grapple with compelling issues and change your mind about things you thought to be permanently engraved in your world view. This is not so much giving way to some other person's point of view, but mutually adopting a new paradigm that is different, but inclusive. Any true step forward includes all steps taken in the past, and this is a time when you can truly understand just what that means.

Pluto in the Tenth House

The world of career struggle and striving to be on top of the heap is often characterized as "a real war out there." That doesn't have to be, but it will during this cycle unless you do what you can to defuse volatile situations that may arise. Challenges to how you earn your living and to your reputation may force you into adjusting to change and devising new directions that are more profitable and satisfying than the old. The rule of thumb is this: If it fails, don't linger on it, move on. Your freedom of action is only restricted by the deadweight you are willing to carry, so toss it overboard like Jonah, because it's all for the best. This may be the opportunity you have been waiting for to reinvent yourself, or at least be seen as doing so.

Pluto in the Eleventh House

Time spent in good company can be transforming or traumatic during this cycle—you decide. When those closest to you have personal issues, understanding and acceptance are the preferred path, while conflict and confrontation are to be avoided. In an atmosphere that seemed easy, you will find out just how individual we all are and that imagined threats can seem very real until they are exposed to be without substance. This is what the Tower of Babel was about, but also the United Nations. When we suddenly all appear to be speaking a different language, it's not time to flee, just time to hire a translator. Relearning what being close means sometimes requires finding a new vocabulary. That's the way you make old friends new again.

Pluto in the Twelfth House

Like the rumblings of a half-dormant volcano, your interior world may be calling to you. This is a good time to dig deep and find out just what's bothering you and bring it to the surface. Hidden agendas, especially ones hidden from yourself, can block your path without betraying much of a clue, until you bring them into the sunlight and resolve real needs long neglected. On a more mundane level, power plays are engaged behind the scenes, and mutually smiling faces may be secretly at war, so take care you don't get caught in the crossfire when it breaks out. Tend to your own issues; stay out of other people's issues and they'll stay out of yours.

NORTH NODE IN THE HOUSES

North Node in the First House

As far as responsibilities go, the buck stops with you this month, and you will take an extra share of personal credit or blame, depending on how you deport yourself. Be careful with new commitments because they could turn out to last longer than you had anticipated, especially where personal relationships are concerned. In that vein, you may find yourself giving your partner a bit of a free ride when it's your turn to pick up the tab where just the two of you are involved. Be generous, as anything you give out now will come back with interest later—what goes around, comes around. Just don't bite off more than you can chew. You don't need to be extravagant to be appreciated.

North Node in the Second House

New financial responsibilities may come your way this month that require you to make extra outlays but may also serve to boost your credit as well. It's important to check twice before signing on the dotted line, because it will be easy to get yourself in deeper than you suspect. Read the fine print, and avoid okaying automatic charges that mount up behind your back. Found money, in the form of an unexpected gift or legacy, could also be in the cards, as good deeds you have done in earlier times get a long-deserved reward. Similarly, old debts may finally get repaid and former projects pay off dividends in the form of royalties or revived interest. Collect your piece of the action.

North Node in the Third House

Things you say to people this month could have in them the makings of long-term commitments, so watch what you say. Others may read more into what you say than you intended and think they've got a deal when they don't—or at least not when you meant to make one. To avoid disappointments and unwanted entanglements, choose your words carefully and remember that handshake agreements often hold as much water as a signed legal document. On the other hand, be on the lookout for especially good new ideas coming your way that could have unanticipated results. Thinking big will place you where you want to be, whereas thinking small will more likely result in entanglements. The devil, not the gold, is in the details this month.

North Node in the Fourth House

Responsibilities at home may take up more time than usual this week, so set some time aside. Also, parse your schedule so that you don't take on too much because of the extra duty and that you don't wind up handling more errands and odd jobs than you meant to. Nevertheless, you may find that while your attention is turned inward, nice surprises may be coming your way on the professional front, as your reputation gets a boost, perhaps from something you'd done long ago but had forgotten about. If someone offers you a free ride, by all means take it and don't be modest, because you will deserve the extra attention and weight taken off your back. Collect your rewards.

North Node in the Fifth House

You may find that the price of having fun goes up this month—specifically, because you will have to make some commitments in order to get what you want. That applies particularly in the area of romance, where a quid pro quo situation can arise that may demand more of you than you had anticipated. It may seem worth it in the short run, but how about the long run? You will have to decide, and then back off if you're not ready to go all the way. Look for a surprise show of support from close friends and superiors, perhaps your reward for an earlier good deed or a job well done. If you've got kids, expect them to hit you up for something—that's what kids are for, after all.

North Node in the Sixth House

New responsibilities in the workplace can tie you down this month, but only to the extent that you let them. Therefore, watch what you say you'll do, because chances are it will entail more than you think. Still, what you take on now could be the kernel of much larger things to come, so it's also an opportunity. Unasked-for favors are likely to happen behind your back, so don't expect to actually see some of the nicest things that happen for you, at least not for a while. When you finally learn about it, you'll be pleasantly surprised. Watch out for other paybacks that might not be so agreeable, and don't be blindsided by those who do not have your best interests in mind.

North Node in the Seventh House

If your partner offers some spontaneous payback this month, don't be surprised and don't turn it down. Let some of the burden be taken off your shoulders and allow others to pick up the tab once in a while. Giving can be better than receiving, so give them the chance to enjoy it. Remember not to tie down a partner more than they really mean to, as the possibility for overcommitment is afoot and you don't want anybody to back out later because they didn't really mean it. Nevertheless, long-range deals and partnerships that deeply affect you are favored right now, particularly where you stand to bring in more than you put out. Don't look a gift horse in the mouth, just get on and ride.

North Node in the Eighth House

You may find that others are particularly inclined to give you credit this month, financially or otherwise. In fact, they may be willing to take on some responsibilities that become a direct payoff for you and even shoulder some of your debt burden. A surprise check in the mail or some other unanticipated financial blessing may be just around the corner—count it as gravy when it comes, and write it off to good deeds you must have done but forgot about. There may be a tendency not to let go of things right now, but moving out the old to make room for the new may be a necessity, and you won't regret it. But recycling things can be an even better alternative.

North Node in the Ninth House

New ideas and approaches to handling large-scale matters can get you personally involved in undertaking new responsibilities you might have avoided earlier. A fresh look at an old subject gets you interested again, enough to commit to spending some more time there. Don't get bogged down in details or you'll miss the opportunities the bigger picture has to offer, where you can really make an impact. Keep the phone and e-mail handy, as good news is most likely to reach you through those channels, like messages from old friends bearing gifts, or other rewards from your past life. Accept with grace and thanks, but know that in some way you have earned them. Small good fortunes, oft repeated, can mount up.

North Node in the Tenth House

This may be an important time to throw some serious effort into promoting yourself, taking up new responsibilities, and committing yourself to deals that will further your career. Efforts you begin now can have lasting repercussions down the line, but may require you to put in more at the beginning than you'll immediately get back. Don't take on more than you can comfortably carry, but don't miss out on opportunity, either. You may find more instant rewards on the home front this month, where you luck into something nice or are the recipient of an unexpected gift. Under any circumstances, you will feel especially welcome and tempted to come home early to enjoy the experience.

North Node in the Eleventh House

You may find that this is just the right time to curry favor with higher-ups and those well-placed to assist or invest in you. To succeed, however, you will have to be ready to commit to obligations in return that you should take a closer look at. What comes easily now may weigh you down in the future, so choose your responsibilities with care. Extra dividends come from children, and opportunities to enjoy yourself physically turn up unannounced, almost gift-wrapped. Creative thoughts come to mind, especially those linked to past events, which can come to fruition as you finally put on the finishing touches. Follow your instincts and let the pieces fall together by themselves.

North Node in the Twelfth House

You may find this month that some of your more important commitments may be made behind the scenes and even behind the backs of everyone around you. The need for privacy about sensitive issues will probably be the reason, but it can siphon off some of the energy you would ordinarily use elsewhere. The need for you to be a listening ear and a shoulder to cry on can leave you booked, but mum's the word as far as the rest of the world is concerned until you finish helping other people resolve their personal problems and can then go back and tend to yourself. If there's such a thing as earning really good karma, this is the time for it. Oddly enough, a relatively free ride at work allows you the time to do it all.

Chapter Four
LUNAR RETURN ASCENDANT BY SIGN

Ascendant in Aries

Your first and last impressions of this lunar month will be of a high-speed, action-oriented, aggressive period during which you rev up speed and push the envelope. This is partly because events tend to fall into place and characterize themselves rapidly without having to go through a period of development. Divisions of life seem to suit their own natures and don't need reworking before connecting. It's like you can skip right to the front of the line. That also gives events the look of being in your face and demanding instant attention. Follow-through, however, may be lacking, so it can be wise to conserve your energy and temper your efforts so you'll have steam left for the home stretch. Nevertheless, this month favors the sprinter.

Ascendant in Taurus

This lunar month gives the outward impression of a bit of a steamroller that, once begun, takes no prisoners to the very end. Irresistible charm mixed with dogged determination is the combination that will see you through and enable you to make the most of it. Concrete solutions are preferred right off the bat, and depth of meaning counts. Subtlety may be put on the backburner in favor of a down-to-earth, practical approach that can sometimes bypass the niceties of the situation. Don't stand on ceremony, and expect the rest of the world not to, either. When you call a spade a spade, you're in like

Flynn, but beat around the bush and you'll be relegated to bush country. This is a down-home month, which favors the long-distance runner.

Ascendant in Gemini

This month comes on gently, replete with subtleties and options to see things another way until you really understand them. Because it's more diffuse in nature, getting down to brass tacks may be frustratingly hard because of multiple options, too many choices. The power of reason, however, is in the ascendant, and you can always talk your way through any situation if need be. You will need to avoid prevarication, however, or chaos will result. Stick to the truth, even when it seems multifaceted. Action takes place on multiple levels, so be ready to pat your head and rub your stomach at the same time. Many winds, often inconstant, caress the landscape, so this month favors the diversified mind.

Ascendant in Cancer

This is a month to either lay siege to or take by storm. Halfway measures won't do the trick. The tendency to either hole up and avoid confrontation or come out swinging may alternate and thus confuse. At times, you may be best off cajoling your way into the action or inserting yourself by stealth like a Trojan horse. A mixture of caution and impetuousness is the middle road, however, and you'll need to feel your way along while appearing to stride. Paper tigers abound, and the seemingly meek may have the sharpest teeth, but you will meet success if your actions remain consistent with your feelings. That means, if it feels right, go for it, but if it doesn't, back off until there's a sincere opening. The operant phrase for the month: peace through strength.

Ascendant in Leo

This month is ready to befriend everybody, a real hale-fellow-well-met. It will offer a good time first, and look into things later, but it is the later that you have to watch out for. Don't forget to bring a pair of hip boots, considering what you may have to wade through, but also bring your party gear, as any occasion can be an excuse for a toast. Life is bigger than life; it's damn the details, full speed ahead. It's the little things that can trip you up, however, so don't let your enthusiasm overwhelm your good sense in the middle of the fray. When you see behind the smiles and handshakes, you can then get down to the business at hand and make the most of the moment, using the expansive energies to help roll you along. The best path: progress through joy.

Ascendant in Virgo

It may seem like you have to have permission to start anything this lunar month, but once you get a handle on the protocol required, you can get things moving. Events will either be handled just so, or they will be chaotic and sloppy, and not much in between. Whenever you can get a written agenda ahead of time, you'll be ahead of the game by knowing the rules. Once agreed upon, however, this approach can quickly see you through the complexities of the month, as it's all connected, a part of the plan. Keep your eyes on the path right in front of you, and the road will be clear. Omit needless words, edit your thoughts, and you'll be on the path. Success comes through structure.

Ascendant in Libra

This month will be characterized by leaps to establish order followed by shifting things around, and then shifting things around some more, a balancing act that continues until the end. This is not necessarily because there is not order in the first place, but just to make sure. It's kind of like rearranging the living room furniture ten times until you decide you're happy with the first arrangement. Nevertheless, you will be working for a good cause, and that is harmony, beauty, and fairness where judgment only comes after due deliberation. Expect bouts of arbitration, compromise, and opposing sides agreeing to meet halfway, or at least trying to do so. A month is not long enough to really do the job! The task this month: striking the best balance.

Ascendant in Scorpio

You may never find out exactly what this month was about, because it is characterized by forcefulness and deeply hidden agendas that elude exposure. You can get your way by sheer might, but one power play will evoke another, so patience is the better path. Profound change is underway, but not on the surface, so retreat and avoidance may be the recommended strategy until the change is over. What you can't overcome, don't try to. Back off and it will move on by itself. The essence of transformation is not to fight it, but to allow it to happen. If the best you can hope for is a Pyrrhic victory, sound retreat with honor. Eschew sacrifice, conserve resources, and await change.

Ascendant in Sagittarius

Broad visions, grand illusions—it may be difficult to tell them apart this month, which is characterized by expansive energy and enthusiastic vigor that may at times make you a little less than wary. The trick is not to bite off more than you can chew, as there will be plenty of details that demand attention that could be overrun by an all-encompassing, one-size-fits-all approach. Harness the excitement, channel the passion, focus the fervor, and you'll be in command of a juggernaut. Allow this horse to simply have its head, and you'll wind up in the back pasture. This is a month that begs you to take the reins, but not too tight, and ride it like the wind until you capture the prize and wear the laurel. The picture: controlled thunder.

Ascendant in Capricorn

This month reeks of self-control, focused intentions, and channeled energy. It does not burst upon the scene, but arrives with reserve, decorum, and, above all, no frills. Don't waste time shilly-shallying, but get right to work and keep at it until the job is done. Efficiency and effectiveness conserve precious energy, and a few words are better than a paragraph. You will need to respect a strong sense of proper approach and established methods in order to get the best out of circumstances, and playboys need not apply. The reward is in a job well-done, without stopping to smell the roses. It's the destination that designs the journey and not the other way around. Be goal-oriented, play to win, and take no prisoners.

Ascendant in Aquarius

This month proceeds in all directions at once, but it escapes the chaos that could occur. Rather, by pursuing each individual path, the totality comes together and is accomplished. Tend each individual tree, and when you stand back, you'll have a forest. Take a pluralist approach, which allows everybody to be right as long as it promotes the general good. Eat bread, not cake, and keep your head on your shoulders. The common touch is the Midas touch right now, so don't put on airs or pretenses of superiority or exclusivity. Prima donnas get voted off the island. It's one for all and all for one—if it's good enough for anyone, it's supposed to be good enough for you. Expect excellence to suffer somewhat, but fairness to blossom. The popular vote rules.

Ascendant in Pisces

This month has a way of eluding definition, and if you try to pin things down too precisely, they just slip right away into the mist. Intuitive approaches succeed where structure fails, and flying by the seat of your pants may get you where you want to go faster than using a flight plan. Faith and hope are the pillars of wisdom for a time, and don't ask for proof up front, as it's likely to be an illusion. Doubting Thomases regret their actions, but the possibilities of sainthood beckon. Follow your still, small voice and you'll float through the month unscathed. Images you project are as real as anything else, which lets you rule with your imagination and chase your dreams. The command: sail on through the mists of night until morning.

Chapter Five

LIGHTS AND PLANETS IN ASPECT IN THE LUNAR RETURN

SUN ASPECTS

Sun Conjunct Moon

Your lunar month starts out with a rousing New Moon and keeps the feeling of newness all month long. It is also your yearly lunar high, one of three major energy peaks during the year. Your action and reaction instincts blend into one to enable heightened aware-ness and quick-footedness. The tendency to shoot from the hip will be rewarded with more bull's-eyes than you would expect, but the tendency to see issues only from one side does not help diplomacy. It's likely all your eggs will be in one basket this month, and if not, you'll probably put them there. Show a unified front, but let others have a piece of the action.

Sun Opposite Moon

Because this month starts out with your yearly lunar low, a personal full Moon, you'll find that primary energies are very split and if you try to focus on just one thing, dis-traction results. Thus, diversify and let your energies go where they are called. It will also mean depending on partnerships and sharing to put together all that is necessary for your needs. The challenge will be in just how and when to pass the ball to each other to avoid self-defeating pushing and pulling. You may have to work harder for less immediate gain, but the spinoffs will be worth it.

Sun Sextile Moon

Cooperation is enabled this month by the energy of the Lights flowing nicely and evenly together. Conservation of energy will come naturally and it will be more obvious than usual how to work hand in hand. With the Sun reinforcing your natal Moon, you'll feel more at ease with both men and women, and social tension will tend to abate when you are around this month. Success this month may be measured more in regular steps than great leaps, as regular methods and channels will work well enough without having to push beyond them, though that might not always have been the case. And, when things don't work out the way you imagined, you're more likely to see the bright side of it and make the best of the situation.

Sun Trine Moon

The feeling that you can get away with anything may be more accurate than you think this month, as energy flows easily in leaps and bounds, and obstacles that might ordinarily be in the way are either swept aside or not there to begin with. The Sun reinforcing your natal Moon lends you the quiet emotional confidence that you are up to the task and that, whatever the details, your basic approach is correct and will be supported by other people and events. If you push the envelope, it will come naturally in the course of things and not likely cause problems or dispute. The only drawback might be that you assume too much, get too lazy, and forget to give that extra push to keep on course.

Sun Square Moon

Conflicting but confluent energy streams make this month a challenge for you to keep in focus and act effectively. The month's general direction can rub you the wrong way emotionally and goad you into actions that waste more than they gain and leave all involved tired and frustrated. This is the time to learn the lesson of not throwing good money after bad—don't go that extra mile trying to fix a situation that either doesn't respond or doesn't really require it. Agree to disagree and move on to a place where all can work together. This is as much a part of your current inner mood as external circumstances, but this, too, will pass. When priorities compete, no one wins, so choose a third option.

Sun Conjunct Mercury

Forceful ideas may be the flavor of the month, but remember that they are a two-edged sword. Enthusiastically and well-supported opinions win friends, but insisting on one person's opinion does quite the opposite. The danger to watch out for is the inflated ego having more weight than the intellect and also a rush to judgment settling issues before all the relevant information is in. Therefore, gather your wits before you launch your propositions, and when you meet with resistance, give it a fair hearing before steamrolling over it. It's easy to be too close to the subject to really see the whole story and find yourself in the pocket of the powerful without realizing it.

Sun Conjunct Venus

When in doubt, this month, turn on the charm. You may be able to get by with anything as you can utilize the force of your own charisma to overcome almost any obstacle. The only difficulty you may run into with this approach is that though your faults may get overlooked now, they may become apparent later, so you have to be able to deliver what you seem to promise. The whole month has a tinge of this lovely but risky illusion, coming from others as well, so take care not to be overrun with momentary desires that may bring longer than intended consequences. Armed with that caveat, however, you can look forward to a positive spin on circumstances whenever you choose it.

Sun Conjunct Mars

Ambition and leadership join hands this month and great strides can be made in pushing an agenda to its conclusion, because the amount of energy and aggression to support it will be relatively boundless. Kept within proper bounds, that can lead to success and achievement. Pushed to excess, this tendency can backfire, if in your enthusiasm and excitement you roll over the opinions or desires of others without due consideration. If you win a victory, do not push on to further damage the opposition, as that wastes energy and shows lack of restraint, neither of which are qualities of a leader. Use any leftover get-up-and-go to consolidate gains only, and beware of going a bridge too far.

Sun Opposite Mars

Conflict can be a central theme this month as irresistible forces buck immovable objects and adversaries cross swords simply because they have nothing better to do. The urge will be to take sides and battle it out, or to keep changing sides each time you think you've got a winner, but neither is a particularly useful course. Your greatest friend will be delay right now, like the Roman general Fabius who defeated Hannibal and all his elephants by not showing up to fight. When forces are in conflict, let them wear each other out until they no longer have the strength to oppose you. Then move in and take command. That way you will not only be looked upon as a victor, but as a peacemaker as well.

Sun Sextile Mars

Contained but plentiful motivation and energy are a trend, and you may be able to find ways to make great forward progress through diplomacy and established channels that open up to you and serve your purposes. The ability to find the gap in the line and race through triumphs over the tendency to push and tackle, which allows you to make a lot of yardage. Teamwork triumphs over tempers, and the well-oiled machine proves its worth by passing obstacles rather than confronting them. You may find you don't have to choose your battles, but can simply avoid them and get what you want. By respecting boundaries, doors open up to you that would otherwise be locked. By offering assistance, assistance is gained.

Sun Trine Mars

The energy required to stride across your landscape unopposed is on tap. The accomplishment can be achieved by a surfeit of energy available and by cleverly avoiding or overstepping obstacles that can take up time and sap vigor. This is the time to bypass knots of resistance and leave them to be dealt with later after you've made your maximum impact. Involve others in your quests and use their power to boost your own and clinch the deal, leaving details to be ironed out later. Resist the temptation to take things for granted, as when you have the advantage, it is essential to grab it and secure it, or it will simply retreat from your grasp for lack of attention.

Sun Square Mars

Going a bridge too far and still not getting there is easy to do this month, and the hardest thing to do will be to conserve energy and choose your targets wisely. It will be tempting to grapple with issues after they have been dealt with sufficiently and waste energy beating a dead horse. When it looks like you're getting entangled in a thicket, resist the temptation to try to cut your way through—rather, back off and try another approach. Energy is drained by gearing up for contests that never occur, and equally sapped when you are blindsided and react too late. The best solution is to stay poised on your feet and wary, and strike only when the situation is ripe and can be finished quickly.

Sun Conjunct Jupiter

There is an innate inclination to overdo this month, and although you may find yourself catapulted into new areas of interest and expansion, put on your plate only what you can consume, or you'll consume too much. Broadening your horizons can open up your head, but it can also go to your head, where delusions of grandeur and illusions of affluence promise horns of plenty that run dry too soon. This does not mean opportunity is not knocking, just that it's time to look gift horses in the mouth and judge which one is really fit to ride. Overshooting the mark under the assumption that you just can't fail is the pitfall. Riding this excellent wave is the challenge.

Sun Opposite Jupiter

Projects begun this month may suffer from conflicting goals, overspending, and uncontrolled growth. This will stem not from fundamental misdirection, but because of trying to have it both ways, burning the candle at both ends, and thus depleting available energy and resources. A lot of either/or choices will be presented, often between equally good prospects, but beware choosing both or straddling the fence. It will be better to choose one and let the other go, or save your energy until something more consistent and delineated proffers itself. Charm and hot prospects abound, and you can be one of them, but follow-through that stands the test of time will determine the final outcome.

Sun Sextile Jupiter

The ability to expand your horizons swiftly but with careful measure is the key to success this month. Rapid forward motion and new directions do not require smashing down barriers but rather the clever exploitation of existing channels to bring you to new places. You can sell yourself with a combination of great promise but understated hype. Just the suggestion of great opportunity serves you better than the hard sell, and people will buy what you are promoting based on your ability to unify and develop with minimum risk and maximum profit. Thus, transform small possibilities into great ones and you will get further than if you appear to change the world.

Sun Trine Jupiter

Rampant progress is available if you want it right now, or you can just sit back and watch it all pass by, assuming you've got a piece. You can sell big changes simply by putting them on display, but you will need to throw your personal commitment behind them in order to keep your rightful share. This is because you will be riding trends larger than yourself, and if you attach yourself to them, you'll get the credit. If you don't, others will. Thus, your leadership style must be inclusive, and you must remember that you're just steering the boat, not making it go. Your ability to find the right channel is on the mark, and if you merely point to it, others will follow. Make no assumptions—act on what you observe.

Sun Square Jupiter

Promising situations can snowball into unmanageable monsters this month unless you are watchful and trim ship regularly. The temptation is to replace the old with the new to the point of diminishing returns, so choose carefully when you have gone just far enough and then move on. Overdevelopment leads to waste and slowing progress across the board, mainly because prospects beckon like the sirens to Ulysses' ears. Don't throw all your eggs into one basket, however good it looks. Be willing to be held back by colleagues. Less can be more and bigger is not better, once you have made your initial innovation. You will get as much credit for what you didn't do as what you did, in the long run.

Sun Conjunct Saturn

Slow and steady wins the race right now, even though you may feel held back or overworked and underrewarded. Projects undertaken now should be long-range and not accepted lightly, but only after due consideration of everything involved. Efficiency and conservation of action are essential, making a single good move worth more than a dozen mediocre ones. Do not suffer fools lightly, as you haven't time to waste, only time to use. It may make you look too serious and severe, but when resources are in short supply, those who make the most of them come out in front. Perseverance is of the essence and rest is at a premium, but the sure knowledge of the goal provides the motivation.

Sun Opposite Saturn

Don't be surprised to find obstacles thrown into your path that slow you down—that's the order of the day, and it's up to you to learn the lessons they give and avoid them in the future. It's one step forward, one step back, and you may feel like you're treading water. You can only overcome your limitations by embracing them, and the final result will be that you find yourself a valued teacher who has seen it all, come through it, and knows the ropes. Accept criticism with grace, especially when well-meant, and hone your skills one step at a time to make sure there are no weak spots. If your foundation is going to be rock and not sand, now is the time to work on it.

Sun Sextile Saturn

Progress through evolution may characterize you this month, and small gains in solidifying your support base will mount up if you have patience and keep at it. You will go further in the long run if you go deeper now and fully explore the implications of what you do before proceeding. A conservative but progressive approach brings progress in increments and has a smaller likelihood of being reversed once underway. You will also find you get more support from others who will see you as less threatening, more reliable, and ultimately someone they will want to invest in. As the I Ching says, this is the time of the preponderance of the small. Steady on the helm.

Sun Trine Saturn

If you act like an authority right now, chances are others will agree, bow to your greater knowledge, and be willing to work more closely with you because of it. This comes equally from your real abilities and experience and your ability to exploit and consolidate them. Aside from appearing more commanding, you will find new support from other experts who will be more willing to welcome you into their fold. This is a good opportunity to establish yourself as a teacher and use your collected wisdom to help others take strides in their own lives. This is both a time to collect on your earnings and reinvest them with care. You have a strong platform from which to launch what you choose—conserve it.

Sun Square Saturn

It is easy to get mired down bucking authority or to get tangled in red tape or out-of-date regulations. It's not a good time to fight city hall, as that will just drain you, but ideal for finding out how to beat the system, probing its weaknesses, and finding its strengths. The same applies to your own inner authority—it's time to find out where you are lacking or insecure and make changes to shore up your self-image and reputation. Like all housekeeping, it is tiring but necessary, and forward motion in other areas may have to be put on the backburner until you have made repairs. Avoid unnecessary risks, and bet on a sure thing if you must bet at all.

Sun Conjunct Uranus

You are likely to want to blast your way through to get what you are quite sure is right, even if you're the only one who thinks so. The chances are that you're right, but also that there may be better ways of promoting yourself that are easier to swallow. You can be blunt but not offensive, sharp but not wounding, shocking but not insulting. Brand new, revolutionary approaches are more easily accepted when coated with honey, not shoved down the throat. Still, original thinking is the order of the day, along with the energy and enthusiasm to support it. Fools rush in, however, so look both ways before crossing the street—the tendency to be self-absorbed and regret it is avoidable.

Sun Opposite Uranus

The psychic landscape may seem a bit jagged right now, with energy coming in fits and starts and follow-through interrupted by bright ideas that demand attention. It may be all you can do to avoid discord and confrontation when tempers flare and thwarted trajectories cause pileups. Impatience and short-sighted arrogance are your own worst enemies, and however right anybody feels, compromise must co-opt conflict or the best hopes will be dashed. Two rights can make a wrong, and this is a likely time to learn that lesson. Instead, distill the best, be willing to toss out pet theories, and use the general desire for progress to actually succeed at providing some.

Sun Sextile Uranus

This is a great time for promoting new ideas, developing exciting research, and finding support for original thinking. The essence of success is to sell what you are promoting as progressive but not radical, needed change but not revolution. By proceeding stepwise, you can preserve the connection of the new (the future) with the old (the past) and allow one to gracefully transform into the other. You will encounter surprisingly little resistance using this approach, and expected roadblocks will likely fade away before your

"Open sesame." Being nonthreatening is the door to success, as is disguising progress in the clothes of the status quo. You catch more flies with honey.

Sun Trine Uranus

If you portray yourself as a font of discovery, you may not be far off the mark this month. Latching onto new trends is the opportunity of the day, and if you are the first to embrace insight, it will become yours to reap the reward. Your only failure would be to consider it obvious and have another wind up getting the credit. Given the opportunity, you can be the life of the party, sparkling with wit and wisdom. You should not cast your pearls before swine, but take the opportunity to get them to the bank while they're still brimming with value. The brilliant mind is often a lazy one, and you've got to add perspiration to that inspiration if you are to cash in on it.

Sun Square Uranus

Truth has many facets and great inventions pop up in different forms at the same time. Sorting them out is the problem, and this is a time when competing insights can lead to conflict. It is easy to feel you are the only one in the right to the point of shooting yourself in the foot, so when challenged, avoid knee-jerk reactions and try out self-criticism to preempt attack. If you have to fight about it, there is probably a shared kernel of truth that will work for all. A shared ride is better than a wrecked car. Push the envelope, but don't break it. The possibility of accidents increases the more you try to pursue all at once, so turn off your favorite tunes and keep your eye on the road.

Sun Conjunct Neptune

An urgency to transmit your personal vision may be enveloped in an inability to get it across clearly. That is, in part, because what is important is not hard facts, but feelings, and they are always difficult to pin down. Therefore it is a good time to stay away from situations where you are expected to be snappy and articulate and gravitate toward environments where you can share emotion, express creativity, and communicate through the spirit rather than the mind. The arts, spiritual endeavors, and exploring your dreams will pay off more than concrete pursuits. Avoid low-level chemical shortcuts, as they will fog your vision and allow illusions to insert themselves into your life as if they were the real thing.

Sun Opposite Neptune

Stumbling blocks come less from the outside than in the form of uncertain direction, confused thinking, and false perception. Progress can be slow because you find yourself interrupted by visions of something better or because you must trample on some of your illusions in order to get ahead. Under any conditions, separating what has substance and what is ephemeral is likely to be a front-burner issue, but in the end, clearing the air will have been a good thing. Thus, for the moment avoid decisions that require precise thinking and take time out to uncover situations where others pull the wool over your ideas or where you are, in fact, kidding yourself. The dreams you can bring to earth, you can keep.

Sun Sextile Neptune

There are times when being slightly vague about what you're doing does not pay off, but this is not one of them. You may find that embroidering the truth with appealing imagination will be just what it takes to get others to accept what they might otherwise refuse. Sugarcoat the pill, then go into pharmaceuticals. In a similar vein, just as others are likely to trust your feelings about the situation, it works the other way around, and the more you leave others to their beliefs and illusions, or use them to your own ends, the happier everybody will be. Burst no one's bubble, and you're in a bubble bath. Come completely clean, and they'll pull the plug. Don't clarify—satisfy.

Sun Trine Neptune

You can be everybody's dream, including your own, and if you take care to pursue them, love and success await you this month. Mainly, that's because you inspire belief and faith and that's what people want to buy. That seems like a dream world, but that's just what it can be at its worst if you don't seize the moment and take advantage to the max. If you just relax and enjoy it, it will pass into a vision of the past, so you have to get with the program and make hay while the sun shines. When you inspire faith in others, you have to prove worthy of it, and though dreaming is free, being a dream is a responsibility. That's why the Beatles broke up. Are you sure you want to be a superstar?

Sun Square Neptune

If you can decide not to decide this month, do it. You may find that, through no fault of your own, much of what seems particularly promising right now will turn out to be illusory, however good it looks at the moment. This is partially because of your own approach in projecting your own fantasies on the situation and partially just the way things are right now. Escapism is just the right escape as long as you realize that's what you're doing. Otherwise, spend some time trimming your visions and cutting through the surrounding mist until you've made your way into clearer surroundings and can commit to a sound course. When you're becalmed, it's time to relax and enjoy until the wind picks up.

Sun Conjunct Pluto

The energy fuel this month has an aura of contained force about it that may be hard to control at times. That's because it is either out of sight and therefore elusive, or an intentional torrent that when released breaks down all opposition and cannot be resisted. That will likely give you the choice of changing or being changed, depending on how the force is directed. On the positive side, you can roll over barriers that were blocking progress, internally or externally, or conversely you can use it to have your way at other people's expense. In some situations you may be forced to do the latter, but the former is preferable by all accounts. Always imagine what it would be like if the shoe was on the other foot, and you'll make the right decisions.

Sun Opposite Pluto

If you encounter persistent resistance this month, back off. Busting through unyielding barriers will get you nowhere, so simply go around the obstacle. There may be paranoia in the air, so avoid catching it yourself and let others be the ones to suffer unfounded fear or suspicion, and let them do it by themselves. Halting progress may result, but better that than involvement in pointless imbroglios that sap your resources without making any headway. For this reason, avoid arguments, however tempted you may be to correct someone who is grossly wrong. The more wrong they are, the more they are likely to fight about it. Instead, sow your seeds on more fertile ground for the benefit of all.

Sun Sextile Pluto

You can make your point forcefully this month and push your agenda through if you are circumspect about it. The trick will be to get others to think they are the driving wheels, when in fact you are. This by itself will be a transformative process that translates raw power into deep persuasion, coercion into cooperative enlightenment. When you identify what absolutely must be done, disciples will pick up the ball and carry it for you in the total belief that nothing out of the ordinary is going on. Throughout, you will get your way by spurring people on, not by rolling over them. An enthusiastic avalanche will follow from a well-directed downhill snowball. Roll it, and then stand back.

Sun Trine Pluto

The more you flaunt your advantage, the more others are likely to rally around you and swell your ranks of support. You've got the power to move mountains by harnessing the raw elements of human need and channeling them to your will, but you will have to be willing to throw your full weight into the effort. It won't happen by itself; this opportunity will pass you by if you don't seize it, so gird up your loins. Power atrophies if it is not used, and you can't dive deep if you don't get into the pool. Flex your muscles while you have the chance, and the chances will continue to appear. When the tide is with you, don't just float on it, use it to propel you where you want to go.

Sun Square Pluto

It can be very easy to find yourself in a jousting tournament where everyone is determined to win but everyone simply winds up exhausted with no gain. There are powerful energies at cross-purposes here, but not in a way that provides victory or defeat, only waste at the expense of good will. So, stay out of the fray. A battle avoided is a battle won right now, and those who hold themselves in reserve will be in a better position to move when the rest are played out. If there is another lesson here, it is to look at yourself and ask where your own anger lies that might push you into taking rash action. When you understand that, you may find that the one roiling the waves is you—and that's easily fixed.

Sun Conjunct Ascendant

A rising Sun purports just what it sounds like—a burst of long-awaited light. Barring other debilitating aspects, this month's driving energy is up-front, in your face, and eminently available to take a ride on. It encourages you to come out of your shell and shine, make a splashy appearance in person, and be a dominating personality in the room. It is great for selling and being a salesperson, as you have a greater power to impress and convince by sheer personality alone. In this respect, as you go, so goes the rest of the world—which will open up in return and let out feelings and thoughts that would have been withheld under other circumstances. Rising Sun, rising star.

Sun Opposite Ascendant

This setting Sun represents a willing handoff to your partner and to coequals in general. Others will tend to be more willing to take up your torch and carry it for a while, and you'll be willing to let them, or you should be. The beginning of evening represents a time when you can relax from your daily burden, have a good rest, and let the night watch take over until you're refreshed. You have the opportunity not only to recharge your batteries, but to develop your personal relationships further. A kind of electric flow is more possible, and mutual love and recognition are more likely, along with a better understanding of the roles each of you play in turn.

Sun Sextile Ascendant

Quiet confidence and charm are the first and likely the last approach to take this month, as the more you allow yourself to fit gently into the environment, the more your surroundings will shore you up and carry you forward. Don't push, just let it flow. Pick familiar ways to express yourself and your ease will be contagious. It will be easier to get things done by just assuming they're in process in the usual fashion. Don't try to alter the course of the river, just bend it. Pick your greatest talents and stick with them, and they will pay off in spades. Compliment others for their talents, and they will hop on board and turn them to your advantage.

Sun Trine Ascendant

If you act as if everything is going your way, it will. That's because you're looking good to others and the only way you could spoil it would be to show a lack of self-confidence. So if you don't know it, don't blow it—proceed as if everything is under control, because for the moment it more than likely is. You can run the show from the background, but don't retreat too far or make the mistake of letting it pass you by. If you want something, ask for it, and you're more likely to get the support you need to get it. If you don't ask, someone else will who may not do as good a job as you could do with less exertion. It takes some effort to make things look effortless, but not that much.

Sun Square Ascendant

You may find your personal life at cross-purposes, where home life or career are in conflict with your self-image and people mistake who you are and what you do based on their own previous projections. You may have to be prepared to explain yourself about the simplest things and still have trouble getting your point across. Similarly, you can find yourself going a step too far in crossing other people's boundaries and might do well to handle sensitive issues with kid gloves so no one comes out with a bruised ego. On the other hand, this aspect opens up possibilities for probing your own weak spots and refining your strengths so they don't get in each other's way. Separate and solve; don't mix.

Sun Conjunct Midheaven

The Sun at midday spreads its widest beams abroad and can be seen from afar, like a flag at the masthead. This month's energy therefore is great for letting people know where you're at. It's great for self-promotion in general, not so much for personal appearances, but for reputation, what people hear about you. Thus, an emphasis on career and your standing in the professional community is easily pursued and much good can come of it. Under any circumstances, the word will be out, and it will be a lot better if you are the one spreading it so you can keep control of what is said instead of just letting the rumor mill grind away. It's a sunny month, so get out and bask in it.

Sun Opposite Midheaven

Around midnight, thoughts turn inward and away from everyday pursuits, and this month will follow suit with an emphasis on the personal side of things. Although career matters may sputter along, that is partially because you will be attending to the fundaments that make up the inner person, which eventually is applied to the race for success in the outer world. Repairing your foundations and tending to your home front are what give you the launching pad from which to catapult yourself into the outer world. Don't think of it as interference; think of it as support critical to your future. Let the rays of the Sun shine on your inner world and charge your batteries for the next step of your journey.

Sun Sextile Ascendant

A focus on ease of communication and incremental and effortless progress is the theme. You'll be looking good and will be able to promote your personal image best by moving along accepted channels and not rocking the boat. What you have spent some time developing can be brought to fruition as people absorb and digest it as something both familiar and positive. New directions are only favored if you cloak them as simply the old ways in new clothing. A good flow of communication and lots of contact with friends is essential and easy to accomplish, and you are carried forward by continual reference to and support of others. Don't go solo, because networking will carry the day.

Sun Trine Midheaven

A strong and easy flow of energy is especially supportive of career moves, and this may be the perfect opportunity to make large strides forward with ease. People are believing the best they've heard about you and forgetting your failures, so look to posit yourself as an expert in your field. Say it, and they'll believe it, and it will further confirm your support. If you get lazy and don't take advantage, you'll bask in praise for the moment, but it will only be momentary. You don't need to be overly aggressive, just confirm that you are as good as you are rumored to be. The impression is already made, so make sure it is not easily forgotten. What is said now will protect you later.

Sun Square Midheaven

Personality issues can really get in the way of advancement this month, with not only your own image but that of a partner conflicting with what is generally believed. The temptation is to strive intensely to make things clearer, but that may only muddy the waters further and manufacture conflict where before there was only misunderstanding. Since egos and the necessity to stroke them have favored odds, go for clearing up issues on a personal level before promoting yourself abroad. The latter is dependent on the former, so get your priorities in the right order. When you tend to the source of rumors, a good reputation follows, along with advancement of your career.

MOON ASPECTS

Moon Conjunct Mercury

You should be able (and inclined) to articulate more clearly the way you feel this month. Making your feelings abundantly clear, however, can be a help or a hindrance, depending on what your feelings are, so watch what you say. It's always good to know better what you feel, but not always good for others to have that information. This period also favors quick and accurate reactions to fast-moving situations, as you will likely figure them out more swiftly and be inclined to move on them more rapidly. Conversely, you will also be able to see all sides of an issue with greater clarity, which can actually slow you down as you weigh the greater number of factors you perceive.

Moon Opposite Mercury

You may have some conflict between your head and your heart, or at least what you perceive and how you feel about it. Apparent contradictions result at first as you opt for either one or the other, but ultimately you can reach a synthesis that allows you to proceed in an integrated fashion. The cost, however, can be in the time spent doing it, so leave extra time for decision-making and don't let yourself rush or be pushed to judgment. It is important that you don't take sides too early in a dispute, or you'll be stuck defending positions later that you don't really support. Don't hesitate to eschew party loyalty and remain an independent, so you can wait until true wisdom sets in to make your move.

Moon Sextile Mercury

You can be very good at making your feelings clear even to the point of being glib, but the more trite you are in your expressions, the more effective they will be. It's a good time to express yourself in adages and aphorisms, which, while cliched, can express even the most radical ideas in the guise of orthodox thinking. When people feel comfortable about your insights, they will more likely fall on fertile ground. Plus, you will gain a more economical expression, as one proverb has within it a paragraph of information and associations, so every phrase will have greater impact and give your personal pitch more horsepower. Contrary to the old saying, here familiarity breeds respect.

Moon Trine Mercury

It's a good time to state the obvious, or treat what you are stating as if it is. Words from you will seem like words from the wise this month, so don't discourage your audience from that impression. The fact is, it will be closer than usual to correct, as insights flow easily, so easily that you may take them for granted. You might want to get what you're saying onto paper so you can benefit from it later. Feelings that formerly may have been difficult to express clearly can be resolved, and your reaction timing in social intercourse will reach new levels. Guard against lazy thinking, however, and be sure to get what's on your mind out into the conversation, or it might as well not be there.

Moon Square Mercury

Restlessness, indecision, and conflicting opinions can be troublesome, and you may have some difficulty separating your feelings from the actual facts in the situation. Digging out the truth may take multiple tries, but you'll get there if you try, try again and refrain from opening your mouth before your brain is in gear. The greatest temptation, and the one to be avoided most, is to simply make up a convenient story to gloss things over. That's not just because lying is bad, but because you're not likely to be good at it now, which could get you doubly in Dutch. Watch out for this same potential in others, when you try to untangle unlikely statements.

Moon Conjunct Venus

A major inclination this month will be to like what you see or make what you see into something you like. Good feelings are easy and intense, and it's easy to love and be loved. Your first reaction will be to charm and flatter, and that will get you everywhere, so go ahead and indulge in it. Avoid being a Pollyanna, however, as not everything you feel to be just wonderful will appear entirely so to others. Make sure you can take the rose-colored glasses off when you need to. On the other hand, your ability to be a rosy optometrist to those in need of a better attitude will be appreciated all around. Good feelings are their own reward, but they also bring reward, in both the short and long runs.

Moon Opposite Venus

Problems with how you feel about what you want may need resolving. That is, you may doubt the validity or rightness of what profits you and thus be put into a moral or ethical bind. Eventually, you'll find a way to resolve them so that you'll be fully satisfied, but it may take a number of reversals of field until you get it right. Don't settle for early decisions, but follow through until the end; otherwise you'll have a bad taste in your mouth, whatever you decide. Avoid overdoing it, hone down inappropriate feelings and reactions, and shun extravagance—all of which may be hard to do. The middle of the road is what you are shooting for, but at least pick a lane.

Moon Sextile Venus

Pleasant feelings, smoothly expressed, are the way to go, along with impeccable taste. Don't spring for the latest fashions, but go with a classic that evokes both beauty and familiarity. Getting what you want has a great deal to do with being happy with what you have, so use what's available to create a masterpiece. Cheerfulness and a good attitude succeed even where confrontation is tempting. Furthermore, they are contagious and bring rewards and support back to you to fan the fires of enthusiasm. Don't hesitate to borrow inspiration from others, as imitation is the sincerest form of flattery, and right now it will be appreciated, not resented.

Moon Trine Venus

It can be easy to relax into an expansive mood and revel in the beauty of what surrounds you—or beautify it some more yourself. Both are in order, as your taste will be obvious and enjoyed by everyone, but you'll only hang on to what you've got if you contribute enough to prove worthy of the praise you naturally receive. Enjoy being in touch with your happier, more enjoyable feelings, but share your insights and wisdom with others, as you only get back what you give in the long run. When you've got more, give more, which shouldn't be difficult unless you really fall asleep at the wheel. It wouldn't hurt to spend a little time figuring out just what you're doing right, so you can buttress your strengths down the line.

Moon Square Venus

What you feel and what you want may be somewhat at cross-purposes, and this can result in both over and under reaction. Being true to your feelings may seem to hurt you, while being false to them brings you gain. Neither bring satisfaction, and half-measures are even worse. It may be necessary is to take a closer look at your values, which are really what is at issue, and then follow that still, small voice that is your conscience, wherever it finally leads. Your greatest enemy is self-deception, a major symptom of the unexamined life. What, and whom, you think you want may lose its shine down the line, and you're not shopping for perishables. Pay extra for something that will last.

Moon Conjunct Mars

It is easy to let strong feelings reign and anger lead you into the fray—you'll regret it, so don't. That's easy to say, but hard to follow, and it involves more than just reining in your emotions—it means finding a way to channel them so you can express them and expend the energy behind them in a positive, not negative, fashion. This energy is floating about and can come back to harm you right on the spot, road rage being a shining example. Fortunately, this tends to pass quickly and not interfere in your overall operations unless you indulge in it. A rule of thumb: Count to ten, and if the feeling still doesn't pass, flee. Expend this passionate energy in love, not war.

Moon Opposite Mars

Keeping your emotional level on the level could be a challenge this month. Feelings run strong and then suddenly fade away before you have thoroughly explored them. The result can be half-expressed emotions that leave you and others hanging, waiting for the next shoe to drop. If you just abandon the situation and move on to another, the results will be haphazard, indeed. So it may be wise to arrange to repeat half-finished situations until you can fill in all the details and come to a resolution that completely expresses not only how you feel, but how strong you feel. The difference between passion and passing fancy is that passion continually revisits and refuels the heart.

Moon Sextile Mars

Your ability to whip up excitement this month is decidedly increased if you let yourself become as much a part of the action as a motivator of it. By refamiliarizing others with feelings they had lost touch with, you can blend your enthusiasm with theirs and empower both. Find where there is latent potential, pick up on it, and help channel it to newer and more useful places. Like a skilled therapist, lead on, don't lead; induce, but don't produce. By sharing your positive feelings, others get to know their own better, and vice versa. By focusing on what is shared, the power to reveal and maximize individual feelings is enhanced through safety in numbers. Rivers flow most rapidly through channels, and only occasionally flood. Channel now, flood later.

Moon Trine Mars

Getting your feelings out into the open may seem so easy this month that you probably won't bother, but you should. Your ability to lead by example is an opportunity that, if taken now, can help establish you as a resource for others who need motivation and guidance. Actions that seem obvious to you may not be so at all to another, and this is the time to share that kind of insight. Feelings that need to be acted upon require a steady hand, which you will have for a time—and if you act on them, they will grow and be a support for you. Further, you'll not only know yourself better by learning through experience, you'll gain a reputation for doing just that, which will profit you in the future.

Moon Square Mars

Emotions and actions may be at constant cross-purposes this month, so go out of your way to avoid inadvertently hurting other people's feelings, and don't blame them for saying things they may regret. Rather, take the time to delve deeper into troublesome issues and hammer out a resolution by repeatedly attacking the problems in good faith. Wherever there's an argument, there is room for compromise, as black and white issues are rare, indeed, in the real world. When things start to escalate without improvement, call time out and start again in a slightly different place. When impatience reigns, insert patience. When intolerance appears, provide tolerance. Perseverance triumphs.

Moon Conjunct Jupiter

This month is part of a year-long trend toward optimism, ebullience, and emotional expansiveness that brings out the positive side of every issue. It's not that there aren't things that need fixing, but for once you're likely to find the good in whatever you encounter and thus turn it to your own advantage by having a handle on it. For now, the grass is actually greener on your side of the fence, and if you don't think so, take a second look. It is very much the time to make hay while the Sun shines, as this kind of energy smiles on you only one year out of every twelve. Hoard your happiness and collect its seeds, so there will be plenty to go around in the future.

Moon Opposite Jupiter

This month, and this year, show a general trend toward devising new ways to fill in emotional gaps. This can lead to extravagance and overdoing things in general, so watch your waistline and your wallet. The instinct of generosity is heightened, but so is occasional indiscrimination in its use, which may cause hurt from those who fail to return it. The way to avoid that is to give only where you know true feelings lie—but that is one of the very things that may be elusive right now, so always take a second look. Of course, if you never give in order to get, as you should, this will never be a problem. In any event, it's a time for growing, learning moral lessons through personal experience.

Moon Sextile Jupiter

This is a time for creative cooperation where you can be a motivating force in brightening the world around you. By putting a finger on new ways that make people happy, you become a source of happiness and good feelings. This does not require radical departures from the norm—rather, by reminding yourself and others of the already available joys in the immediate environment, you maximize the return on what you put out. Although misery loves company, joyfulness adores it, and enthusiasm is contagious, so spread it around. You get back what you give and you won't have to wait for it. You can feel good by yourself, but it takes a village to throw a party that lasts all year.

Moon Trine Jupiter

Generosity falls on fertile ground, and giving comes easy when your heart has plenty to give. It's the feeling that counts, and the more inclusive you can be, the greater your life feast. Many of the best things in life are free right now, but you need to sit up and take notice, as it's too easy to enjoy them halfway and then let them pass. When you've got the whole enchilada, slap on the sauce and serve it up for all to enjoy. What you seize today will be tomorrow's treasure. Not only this month, but this whole year has the potential to be a pot of gold, not necessarily for the bank account, but surely for the soul. Your bowl is fuller than you thought, so dig in.

Moon Square Jupiter

Great ideas promise more satisfaction than they deliver, and you'll need to sort out what profits and what doesn't. This is a time to experiment, and expect to try several times before you get it right, so don't blow your wad on the first shot. If you expect less, you'll get more, and this is a month—no, a year—during which you will come to understand just what that means. If you overshoot or undershoot the mark, don't be disappointed, just take another shot. After a while, you'll never miss. Take this period to hone your sense of proportions, improve your timing, and regulate your appetites. You can't find your balance if you've never lost it. Get to know your capabilities better.

Moon Conjunct Saturn

Feelings this year have and will feature greater gravity than usual as Saturn makes its rare appearance over your Moon. Emotions are restricted and less easily or willingly brought to light, but when they are, they are very intense. Heavy-duty attachments are possible, but what is appealing now may be burdensome later, so choose wisely and think long-range, as diving into situations now may mean having to stay there for quite a while. It can be too easy to underrate yourself right now, so watch out that you don't give away or sell for cheap what could be worth a lot more next year at this time. Don't be discouraged if you can't find instant gratification—what's worthwhile is worth waiting for.

Moon Opposite Saturn

Feelings this year have and will feature spurts of greater gravity than usual as Saturn makes its rare appearance opposite your Moon. You may find times that emotions flow easily followed by periods when you just can't get in touch with yourself or others. This can make relationships a sometime thing as you can't always be certain exactly how they are going. Thus, it might be a good idea to put permanent commitments on hold until you determine whether they are as serious as they sometimes seem. It is, however, a good time to throw yourself into your daily work, which could use the extra attention and can do without the extra emotion.

Moon Sextile Saturn

You may find yourself reaching a plateau of emotional stability such that you can be a first-class shoulder to lean on. When others come to you with troubles, your advice will reflect a calm and steady example and a refuge from the storms of life. Although you may find comforting platitudes trite, others won't, and you can breathe new life into old maxims by adding fresh insight and updated applications to them. You'll also find that addressing the problems of others helps you better understand your own character and allows you to rediscover and strengthen those parts of your nature that have stood the test of time particularly well. Self-examination is its own reward.

Moon Trine Saturn

You may have a steadier hand this year than you ever thought you had, and if you let on to it, you will be a shining example to others and will gain their admiration. Patience comes a lot easier and lasts a lot longer, and in the face of external upset you will be unflappable compared to some around you. This may well earn you greater responsibilities and the increased benefits that come from them. Accomplishments will mount through stick-to-itness, and now is the time to put in extra effort while you have the extra stamina. This mood won't be with you forever, however, so now is a good time to examine your work habits and see just how you manage it all. You may need your notes later.

Moon Square Saturn

It can be hard not to be overcritical right now, and there is a lot of negative thinking in the air that you should try to stay clear of. What you mean as constructive comments can be mistaken for derogatory remarks by others, so attempt more than you ordinarily would to put a positive spin on things. That goes for self-criticism as well, and you should probably give yourself a lot more credit than you think you deserve. Now is the time to look for your strong points, not dwell on your flaws. There is a bright side to everything and every cloud has a silver lining, and if you can't find either, just ask someone. Don't sit in the dark—turn on the lights.

Moon Conjunct Uranus

Uranus conjoins your Moon only once in a lifetime, and this is it, so hang on. This period of about a year can be a bumpy ride, full of surprises, shocks, and emotional tours de force that come out of nowhere in an instant and vanish equally as quickly. Avoid putting yourself into risky emotional situations unless you're ready to deal with feelings that can go to extremes. On the other hand, you can find that the changeable weather gives you great insight into your inner workings and self-discovery comes like a thunderclap when you least expect it. Don't pursue situations that grate on your nerves, but opt for areas where you can reinvent yourself.

Moon Opposite Uranus

Uranus opposes your Moon only once in a lifetime, and this is it, so hang on. This period of about a year can be a bumpy ride, full of surprises, shocks, and emotional tours de force that come out of nowhere in an instant and vanish equally as quickly. Fortunately this unpredictable psychic weather comes in patches with calm periods in between for you to recover. Emotional experimentation can uncover new sides of yourself you didn't know were there and inner resources that surprise you. Annoyances abound, and this is the time to learn to brush them off until you're unflappable. Outbreaks of strong feelings rush in, and rush out. What you'd pay for at an amusement park, you're getting for free.

Moon Sextile Uranus

This is part of an extended period of self-discovery where you can take risks with your heart and usually come out the happier for it. You'll have a good sense of just how far to go, how far to push the envelope without it breaking. The new and unusual abound but don't threaten, and new inner landscapes appear that are real eye openers. You also have the ability to be a guide who introduces others to new things in surprising ways while still staying in the comfort zone. Original thinking is the order of the day, and wonders are to be found not only in strange places, where you'd expect them, but lurking in the corners of everyday life, waiting to amuse, entertain, and enlighten.

Moon Trine Uranus

This year is a voyage of discovery laid out like a package tour, where unexpected treasures of the mind and heart spill out easily, without seeming effort. In fact, you may take for granted what others would consider challenging or even far out, so that you may not appreciate the novelty of what comes your way. Try to see yourself as others see you, and you'll have a better idea of how to utilize your gifts both for your own gratification and for the general good. You've got the ability to be a leader with a fresh and original approach, and people will flock to if you just let them know you're there. Treasure your discoveries and share them.

Moon Square Uranus

Off and on this year you may find impatience overcoming you, and it's hard to put your finger on just why. Inspiration comes in flashes, but the more you chase it, the harder it is to catch. Half-resolved issues annoy and sometimes anger when they won't yield to reason, so it's a good time to learn lessons in patience and stability in the face of unpredictable and random events. Eccentric behavior can seem the norm, and you may come to appreciate quirks that once put you off but now appear as unique artifacts of character. You're peppered with original thoughts and bright ideas, but you'll want to test them well before relying on them, as some are incomplete.

Moon Conjunct Neptune

It's a unique year when feelings may be hard to articulate because your inner emotional processes are going through a period of change and redefinition that won't solidify for a while yet. If you seem distracted from time to time, it's because you are and should be as you sort out a variety of personal approaches you hadn't really examined closely before. Since you are in the process of change, it's a good idea not to make snap decisions based on previous assumptions, but rather take extra time to see if you may want to reconsider your approach and take another tack. You may not see the changes going on right now because you are too close to them, but you'll look back later and they will be quite apparent.

Moon Opposite Neptune

This year you may find that your reactions are not always as solid as they might be, interrupted by periods of distraction or reverie. That is actually a sign of the creative process cutting in sporadically, which is to be welcomed, but it can also seem like you lose touch at the same time. A senior moment, as they say. It seems like you are working on two levels at the same time and alternating between them. Although this eventually means doubled progress, while it's happening you may miss a stitch or two unless you doublecheck before you decide to go with spot decisions or immediate reactions. When you think twice, it's all right—just make sure you do.

Moon Sextile Neptune

You are in a fairly lengthy period where you can get extra mileage out of your imagination, and your intuition will be more in tune than usual. It's a time when you can do more than daydream, you can finetune your dreams so that they can become a reality. What's especially unique is that you need not go far afield to find what you are looking for, because it's more or less in your own backyard. Look to the smaller things in life to be bringing you the bigger inspirations, as it is in the microcosm that you see the larger picture reflected. Right now you don't have to chase after your goals, but rather be very still and they will come of themselves.

Moon Trine Neptune

For some time now you can paint your dreams with broad brush strokes, and creativity and an innate sense of what's coming next allow you to be very on top of things, perhaps without even knowing it. In fact, much of the challenge at the moment will be figuring out why you know what you know. Don't take talents for granted but explore them in sufficient detail to be able to teach and set down what you have learned. Thus the realms of your imagination can be sufficiently mapped so that you can get back there should you lose your way later. Similarly, if you can be a guide for others, you can have company to explore your own dreamscapes and bring your fantasies to earth.

Moon Square Neptune

It can be hard to put a finger on your feelings right now, a trend that will be sticking around for a while. If you allow yourself to proceed on previous assumptions, you can be way off base without realizing it, so take time to question your motivations and see if they are leading you where you really want to go. It's a good time to reestablish the difference between fantasy and reality, as the line between them can get a little fuzzy. Chemical solutions will not serve you well right now and are more likely to exacerbate problems rather than solve them. It's time to clear out the cobwebs of your mind and let in some fresh air.

Moon Conjunct Pluto

This is a very transformative year for you, during which you may find yourself changing greatly on the inside, though it may not necessarily show on the outside. Ideally, it can be a period of emotional retreat and revamping so that you reengage with the world with clearer feelings and more measured reactions. The one thing to avoid right now is any attempt to batter down barriers or force your way to what you want. Don't try to move mountains, just go around them. If at first you don't succeed, don't try, try again—try something else. It is better to wait for change than try to enforce it. In case of rain, stay inside until the weather clears—it will, soon enough.

Moon Opposite Pluto

This is a year of some considerable emotional change for you, marked by periods of clear sailing interspersed with stubborn roadblocks that are better circumvented than surmounted. When faced with change in a relationship, you can just go along with it or avoid it entirely. The one thing not to do is to put up resistance or you may get very frustrated, indeed. Expect to run into power plays, but you don't have to play along—at any given time, you can always sit out the dance and wait for a better tune. In the end, self-control is the lesson to learn, and yielding to compulsions is tempting but counterproductive in the end. If you feel you simply have to do it, don't.

Moon Sextile Pluto

You have a large portion of inner confidence and conviction at your command this year, and you demonstrate personal power not by flaunting it but by quietly exuding it. If you feel certain about something, then it is certain, even though you might not be able to articulate a good reason why. If you want change in a situation, you can effect it by steady and unflagging pressure that is not coercive so much as just convincing. If you keep at it, whatever it is, you will eventually have your way, and without having to start a fight about it. By never doubting the rightness of your cause, you lead by example and others will follow along because it's contagious.

Moon Trine Pluto

This year likely finds you in possession of a kind of fearlessness that can get you through the stickiest situations by faith alone. Where others might doubt, you never give it a second thought, and proceed. This does not mean rushing in where angels fear to tread, but it does mean trusting powers greater than yourself and in the situation itself to carry the day when it might seem impossible to accomplish alone. When you aim at a target knowing you are going to hit it, you rarely miss the mark. It is only when the hand or eye loses confidence that failure becomes likely. Consider the future to already be in your possession, and it is.

Moon Square Pluto

This is a year in which you may have to learn to play power games and force may be necessary to get what you want. The trick is using the right amount at the right time in the right cause. It can seem easy to just cut the Gordian knot, but you've got to be careful not to destroy the very thing you want to gain by applying too much pressure. Achieving your goals may take more push than you expect, as you'll find that others push back and then nothing useful is accomplished. Some things are worth fighting for, and some not, and you may well find that the better part of valor will be to disengage and wait for better opportunities to come along.

Moon Conjunct Ascendant

A rising Moon this lunar month makes for a very "in your face" atmosphere that catapults you up front in person. It's good for selling yourself, but a little relentless, and you should take care not to seem pushy or to appear to be monopolizing the situation. Your reaction time should be faster and you can get your licks in before the next person, which can give you an edge over the competition. It may also be easy to become overly emotional and to overreact to some situations. If you wear your heart on your sleeve, just remember that you risk some breakage by doing so, although the appreciation for your honesty may quite make up for it in the long run.

Moon Opposite Ascendant

A setting Moon this lunar month tends to throw the ball into your partner's court and it will take strong cooperation to get things done. Although that may seem to take away some of your initiative, it is time to get the full load off your back and let someone else carry it for a while. Expect emotions to run high in personal affairs, where partners may be a bit oversensitive and take things more seriously than they are intended. Nevertheless, it's the thought—and the honesty—that counts, so don't feel put off by overreactions. Part of the art of friendship is to give your partner enough space to really make a point, without judgment on your part.

Moon Sextile Ascendant

This can be a month of very smooth moves on your part, as your ability to react to circumstances combines well with your social skills and diplomacy. Gentle persuasion is the approach, and others will find you quite convincing when you use the soft sell, not the hard. When you stick to customary avenues of approach, you'll do better than coming in with ideas from left field. By maintaining a strong comfort level, you'll make the quickest and greatest progress. Emotions are best kept in check and on an even keel, though that does not mean you have to repress yourself, just express your feelings in a measured way so as to be persuasive but nonthreatening.

Moon Trine Ascendant

Your easy presence will let you get away with things this month that you might have had to struggle with earlier. By acting as if it's all a done deal and simply taking progress for granted, it will come along all the faster. It merely takes the bearing of authority to be quite convincing as an expert, one who has the situation under control without giving it a second thought. That doesn't mean you don't have to make an effort—just make it appear effortless. When someone hands you the ball, run with it like it was second nature and you expected it all the time. At times you may wonder why you bother, but bother you must or lose golden opportunities.

Moon Square Ascendant

You may find there is some conflict between personal and professional issues this month, which can have you at cross-purposes with yourself. Mixing work with play may not work out well, and you're probably better off keeping the two separate. Don't take the office home with you, and don't let personal entanglements leak over into the workplace, either. You may not be as smooth as usual in extricating yourself from difficult situations where others won't let go and move on but insist on beating a dead horse. A clean break may be what's needed, but it could be hard to accomplish as swiftly as you like. Therefore, make some room in your schedule for unavoidable time-wasting.

Moon Conjunct Midheaven

The Moon high in the heavens gives you considerable visibility this month, and it's just the time for throwing your energies into personal promotion and career advancement. Your sales skills are in gear and whatever product you're pushing can gain a high profile and a broadened reputation, even if that product is just yourself. Be quick to pounce on praise and pass it around while quietly consigning errors to the round file. What people don't know won't hurt you, but your future could spin on what they do know, so get it out there. Be willing to sacrifice some personal time to get done what you need while the opportunity is knocking at your door.

Moon Opposite Midheaven

You may find that this month has you rather preferring to stay home and tend your fires rather than joust in the world arena. Issues arise that need personal tending to, fences require mending, and any number of matters that involve the home front will likely take precedence and be the focus of your efforts. Matters of the heart beckon and make you realize what your true priorities are and what it takes to give them the rank they deserve. The family front takes the front seat for a while, and strong emotions may surface that had been put away, obscured by the common business of the day. It is time to return to the uncommon, what is real, what is most close to you and supports your deepest feelings.

Moon Sextile Midheaven

This month offers not inconsiderable opportunities to promote yourself from offstage, letting others boost your reputation by showing support for your flag. This is not a time for wild, blockbuster reviews, but rather a flow of steady compliments that emphasize how well you fit in with your surroundings and complement them. In many ways, the status quo is your friend as long as you can make progress by leaning on it and appearing to fit in while in fact pulling ahead. Sell yourself as a team player right about now and the team will give you valuable support in return. Don't nail yourself to their agenda, but go along to the extent that it benefits you.

Moon Trine Midheaven

Hard work and tangible investment in career plans can pay off big, as support is easier to find and work itself can fly by if it's producing. Don't be afraid to spend money on a good cause, as what you spend now will have a better than usual chance of bringing back dividends. Similarly, going the extra mile at work can really come back to you in disproportionate returns. Take big gulps, not sips, in whatever you do, and do it with gusto. The easier you make it seem, the easier it gets as long as you don't take things for granted. When you look like a success, people think you are a success—and you are! Lead with a positive attitude.

Moon Square Midheaven

You may find that personal issues can cloud your attempts to better yourself in the workplace, and what is said about you doesn't correctly represent what you really do. What people see is not necessarily what they get and you will have to make sure that your talents are not inadvertently underrated. Similarly, it is easy to confuse what a person does with what that person is, so any and all attempts to keep an appearance of professionalism and avoid getting personal are in order. Right now, work and play don't mix very well, so keep them well apart to avoid misinterpretations or entanglements that can slow down your career progress. Don't spread, or believe, rumors.

MERCURY ASPECTS

Mercury Conjunct Venus

This month is ideal for anything you can put together having to do with the visual arts and design. Painting, photography, film, interior design, architecture, mixed media, and other related fields all will do particularly well for you right now, so see where you can use them to their greatest advantage. This can range from home decorating to advertising promotions, from building a house to just painting one. Similarly, a strong sense of picking just the right phrase to describe what you mean can turn your everyday prose into poetry and beautify the inside as well as the outside of your world. The creative Muse is at your beck and call—don't ignore her.

Mercury Conjunct Mars

Forceful words may fill the air this month, which can work for you or against you. You can use the high energy and tight focus of your mind to blast off new ideas and propel them into implementation. But, should you press your case too hard, it could backfire into arguments and disputes over either your ideas or the speed with which you are pushing them. No matter how right you are, if you push down other people's defenses they will react disagreeably. Therefore, let the weight of your arguments provide their own impetus, while you provide the enthusiasm to support them in such a way that others embrace your direction and don't fight it. Be positive, and avoid arguments.

Mercury Opposite Mars

Intellectual sparks can fly this month, but you will need to pace yourself, as strokes of brilliance are followed by a shortage of energy for follow-through. Arguments can spring up if you push your views too forcefully or if you insist on pursuing ideas that others find fault with. Impulsiveness, albeit filled with fire and enthusiasm, can be your worst enemy as it can make the best ideas seem insufficiently developed and, in a word, green. So before making a presentation, research all possible objections and have rebuttals ready to present in a reasonable and measured manner. Hotheaded confrontations would be a dreadful waste of the strong energy flow this month has for you.

Mercury Sextile Mars

Well thought-out plans smile on you and you can put words into action this month by enlisting the help of acquaintances and generally networking to get things done. Good communications are heightened and agreements easier to reach as you find common ground for progress and the enthusiasm to roll it along. Small steps mount up, and they will get you further than great leaps right now. Serial improvements have the guise of business as usual but come together to make something unusually impressive. The ability to swiftly and surely put all the pieces together in concert with others is heightened for you, and circumstances conspire to give you calm internal weather in which to use it.

Mercury Trine Mars

Anything you can put into words you can probably accomplish this month, as form and function go hand in hand and your shots are clean from backswing to follow-through. You can make things look easier because right now they just are, but there's no need to let on when you can take lots of credit. Don't hog the action, but make sure you are in the middle of it as you have a lot to gain. Well-considered moves gain you praise, but you can't get lazy now just because it seems easy or obvious. You can't get your point across if you don't speak up—especially now that you're likely to meet with agreement on all sides. Speak clearly, act accomplished, and gracefully accept admiration.

Mercury Square Mars

Words and deeds may be at cross-purposes this month as the distribution of energy is uneven and either overshoots or undershoots the mark. Honing your aim will require some real attention, and frustrations caused by misunderstandings or hasty moves can lead to disagreements and upset if you don't make an extra effort to pour oil on the waters to smooth things over. Making it clear what actions you intend to take before taking them is important, so that you don't blindside others and waste your own efforts for lack of general coordination. Don't assume others understand you or where you are going. Like at a stop sign, make clear signals before turning or proceeding on.

Mercury Conjunct Jupiter

The month will likely see a lot of new ideas turning up all over the place, and the challenge will be to get them mapped out clearly and under control. Not every new brainstorm is a keeper, but enough make the grade to call this a very creative and original month. The emphasis is on bigness and newness, with the general thrust outlined but the details not all filled in yet. The devil, however, will be in those details when you come to them, so don't overstep your limits. The first stroke of genius is not always the best.

You will be tempted to let your imagination run away with you—that's okay, as long as you keep a good grip on the difference between blue sky and the real road ahead of you.

Mercury Opposite Jupiter

Brainstorming explodes all around you, but not always in the most orderly way. On-again, off-again inspirations are sometimes hard to keep track of and even harder to follow through on and the Muse is alternately generous and fickle. When in doubt, write it down so you can come back to it later. The greatest risk is latching on to incomplete plans that can fail later because of missing parts. What looks good on the surface may well be fine, or it could be flawed to the point of uselessness. Only time will tell, so give yourself some time while at the same time enjoying the heightened creativity of this period.

Mercury Sextile Jupiter

A strong current of creativity and clear and original thinking runs through the month, but not in the form of blockbuster breakthroughs or revolutionary agendas. Rather, the rivers of thought flow swiftly along their established channels and together they mount up to a well-controlled flood. Cooperation and joint research better describe what's going on rather than single visionary insight. Multiple small discoveries lead to the solutions of big problems, and truth is discovered as an aggregate, not a single whole. Short journeys laid end to end easily circle the globe, and the whole appears when all the parts are in place. Cross-pollination brings the fullest harvest.

Mercury Trine Jupiter

Grand thinking brings great reward, and it will be easier than ever to approach this combination right now. Brilliant ideas, penetrating solutions, and original operation all spring up easily, as if by themselves. In fact, you may totally overlook them because they appear so obvious, so keep your antennae up. When you think you've got your finger on a winner, let it fall from your lips as if it were just another stray opinion, and then watch for the response as you're seen as an underrated genius. Remember, though, that luck is not always with you like this so take advantage of what you've got and put your brand on it. Patent it, copyright it, and trademark it so it stays with you.

Mercury Square Jupiter

Runaway inspirations can confuse as much as amuse, and be on the lookout for half-baked ideas that are clothed as full-course meals. Chasing down details and bringing plans to earth are equally hard to do, and you may have to go an extra mile or two until you have separated the wheat from the chaff. Avoid jumping to conclusions and unnecessary wastefulness, either physical or intellectual. There's a lot of hot air around and you may have to break out your waders to get through it all. Still, in there somewhere are the kernels of some very bright ideas if you trim them down and nurture them. The trouble is getting them to sit still while you perform the operation.

Mercury Conjunct Saturn

Serious thinking and attention to details are the trend right now, and the scientific method is the approach of choice. Economy of thought and word is the ideal, and if you're not understood, don't rephrase—just repeat. To recall Strunk and White: omit needless words, omit needless words, omit needless words. If you don't need it, get rid of it, trim ship. Of course, it's also easy to take yourself too seriously now, so remember that God is a humorist and you are only sometimes in on the joke. It's a good time for getting research done, signing long-range contracts, and building foundations. Patience is the order of the day; slow and steady wins the race.

Mercury Opposite Saturn

Expect delays in even the best-laid plans and leave lots of extra room in your schedule to take up the slack. It can be easy to be pessimistic, but impatience will do you no good, and this, too, will pass if you take the time to work things out. Watch what you sign, as you could be in for more than you think and extra burdens are easy to come by right now. Don't be surprised if the cat's got your tongue and it's not always easy to express exactly what you mean. Take time to rephrase, set the record straight, and then proceed as usual. Delays are worth it only if you get it right in the end.

Mercury Sextile Saturn

Carefully laid and implemented plans can grow to maturity now, as long as you take the time to see it's done right. Step-wise is the rule, but you can pick up every stitch more easily than you thought as the result of doublechecking and sharing responsibility with others. The image of a well-oiled machine comes to mind, a mill that grinds slowly, but exceeding small—and exceeding well. Quality control projects go well now, as do mapping and measuring. Keeping to the speed limit is a boon when the lights are timed, so everything goes like clockwork. Meticulous engineering marks the fine driving machine —drive on.

Mercury Trine Saturn

Realistic decisions for the future are at hand, and a long-range sense of proportion is the order of the day. Building on a strong foundation not only makes a lasting structure, it's easier to do. All the blocks fall into place and there you have it—effortless, but rock solid. Well, not quite that effortless, really, as sustained effort is required along with patience and stamina to build the edifice you envision, but you don't have to strain yourself—just don't drop the ball. When you have done your homework, you always pass the test, and this is where you learn the truth of that. Keep doing what you're doing and you'll pass with flying colors.

Mercury Square Saturn

Negative thinking can be a problem, as can stubbornness right about now. Avoid easy put-downs or cruel comments that seem smart but do more harm than good. Error-trapping is favored, but you have to make sure what you're attacking is really error. The error may be yours. Frustration with repetitious thinking and redundancy is natural, but you can't get everything right the first time. Try, try again, and it will all work out in time. This is a time for critical faculties, not critical people, so pay more attention to the job at hand and less to those who are doing it. Personal remonstrations are not in order, even though others may indulge in them. Be an example.

Mercury Conjunct Uranus

Your mind sparkles with original ideas and radically new ways of seeing things open up in a flash. Actually, these new attitudes may have been gestating in the subconscious for a while, but they seem to spring into the sunlight full-blown, like Athena from the head of Zeus. The revelations that come now may not be practical to implement right away, but they will profoundly affect the directions you take as time goes by. It may be a good idea to have extra time set aside to capture a clear record of what you're thinking right now, as it may be the foundation for your action in the future. When the Muse speaks, don't just listen, write it down!

Mercury Opposite Uranus

Thought processes this month are subject to interruptions and annoyances, as just when you're getting ready to finish what you're doing, something else breaks in and upsets it. What you're actually seeing is the dialectic in action—thesis, antithesis, synthesis—but in somewhat jagged stages. The point will be to have the patience to absorb the disparate input around you and put them all together into something that makes sense and actually works. There are some really important new directions just lying around you half-finished, waiting for you to take advantage. It is for you to identify them, pick up and sort out the pieces, and assemble a combination that really rocks.

Mercury Sextile Uranus

Support for your dreams and schemes is all around you this month, allowing your brighter ideas to see daylight and get into the working mainstream. Support and cooperation are the key, however, as even when it's your brainchild alone, it will take the input and underwriting of others to get it into operation. So, it's time not to just push yourself, but to lead others along into channels that profit you and make them feel a part of your new approach. Make it a general bandwagon that everybody can get on, but over which you have the most control, having designed it. Like a diamond in the rough, it takes cutting and polishing to truly sparkle, and you have to farm some of that out.

Mercury Trine Uranus

This is a good time to shine and make hay with the insights and new perceptions that have come to you lately. Just open your mouth and speak and you'll get credit you may think you don't deserve for original thinking, insight, and imaginative approaches. What seems obvious to you may seem quite clever to others, so keep your eyes open for reactions to what you say. You could be in possession of intellectual gold that can be parleyed into new opportunities, but you'll have to test the waters to determine that. In other words, wake up to what you are saying this month and see if it's waking others up, too. Clever does more than invent—it builds on its inventions.

Mercury Square Uranus

It's easier than usual to have a sharp or sarcastic tongue, and the opportunities for cutting up with a rapier wit abound. You may find that this can cause more trouble and hurt than you think, so it's a good thing if you hold back a bit, for two reasons: First, you may be putting your foot in your mouth, as mental processes are at cross-purposes now. Second, you may create enemies who would have gladly been your friends had you been more gentle with their mistakes. Impatience leads to shallowness, however surface-bright, so follow your feelings a step deeper before letting them out of the box.

Mercury Conjunct Neptune

An air of either confusion or deception surrounds much of what is said this month, so be on the lookout for both. It's a perfect time to pull the wool over someone's eyes, or have the same done to you. You find fuzzy thinking too widespread, but it's sometimes hard to clear it up. On the more ethereal side of life, it is a great time to put your dreams into words, to articulate your fantasies, write poetry and love letters, and come to a deeper understanding of spiritual issues. Read sacred scriptures, write science fiction, pore over ancient mysteries, and commune with the spirit world. Just don't sign on the dotted line or buy a pig in a poke.

Mercury Opposite Neptune

The rumor mill may be working overtime right now, so be sure to doublecheck your sources before you believe anything you hear. Further, ill-considered plans and schemes may abound, themselves based on incorrect information, so make sure you're not part of them. You've got time on your side, if you use it, as flaws surface fairly quickly if you don't swallow the bait right away. It's time to be a wary fish in a muddy pool. Next month the water will clear, so you'll be glad you waited. Reverie, times lost in thought, and speculation do well right now, as long as you don't rush to implement them. A second look will either give form to the dream or make it vanish like mist.

Mercury Sextile Neptune

Selling your dreams can be easy, especially when you describe them as the dreams of those you are selling to. The realms of the imagination find easy support and you can use inner imaging to bring out the best in people right now. Intuitively, you can judge feelings and help share interior emotions that might have been difficult to express by putting a positive, nonthreatening spin on them. Make it a group effort, with everyone getting credit for their part in making it happen, even though you may be the main spinner. This is a good time for creative work in any field, especially the arts, as a generally friendly environment for safe free-association allows all to open up.

Mercury Trine Neptune

Open hand, open heart have the advantage as solid links with the inner world unfold and allow more intimate expressions of love, romance, dreams, and aspirations. The greatest inner wisdom unfolds with the least effort and you may have to check yourself ("Did I say that?") when accidental profundity comes your way. Listen to what you are saying, and take your own advice. Although you may be a bit on the idealistic side, you've got to have a dream to have a dream come true, and now is the time when dreams are well-made and are more than just fantasy or illusion. Check self-criticism at the door; you will have plenty of excuses to doubt yourself later, and with better reason.

Mercury Square Neptune

If there is ever a likely time that falsehoods should abound, this is it. Take everything you hear with a grain (no, a barrel) of salt and always check it out before you make a move. Sometimes you will encounter only outright confusion or misunderstanding, but sometimes it will be outright lies. Forewarned is forearmed, but more difficult may be the temptation to bend the truth yourself in the interest either of expediency or because you don't understand the situation yourself but want to hide it. If you don't know, ask. Ask several times until you're sure. Never let self-deceit draw the veil of ignorance around you when a little more trouble will bring you truth.

Mercury Conjunct Pluto

You may have the opportunity to use the power of your mind like a hammer this month, but be subtle or it will be useless. You can have strong, forceful, deeply held opinions to sell, but the way not to have them welcome is to shove them down someone's throat. If you want your way, let your arguments do the talking and don't make it a battle of wills, even though you could more easily win that way. It's the difference between alliance and conquest, friends and enemies. It is a better time to harness powerful ideas and make them work for you, spread the gospel around, and let the word be heard. True power does not emanate from a single person, but from the place or situation where the larger picture comes to earth and is made available.

Mercury Opposite Pluto

You may find just the perfect thing to fight over, but cool it; it ain't worth it. Not that one or both of you may not be right—but what is wrong is for one of you to roll over the other. Impulsive moments when you blurt out what you feel when perhaps you shouldn't can give away inner motivations that would be better explored in private. Indeed, the stamina necessary to win a fight may not be available, leaving all parties wounded but without a victor. When in doubt, check it out before opening your mouth, however right you think you are. The deeper your feelings, the more personal they are, and thus the less they may apply to another.

Mercury Sextile Pluto

This is a wonderful time for manipulating people without guilt and coming away with having actually done a good deed. You can literally force your ideas across as if they were simply observations of group necessity, and the group involved will feel compelled to go along. Cooperation is the key, and you can't sell just anything—only things that lend themselves to subconscious desire or motivation, as that's where the real sales pitch will be. Avoid playing directly on fear or other negative emotions, but lean on feelings of faith and surety—like, you just know this will work, everybody can feel it. Group psychology is the order of the day, so be on the inside of it.

Mercury Trine Pluto

Faith in your convictions will get you everything this month and it will be the manner in which you can forcefully get across what you deeply feel to be the right thing that will have others following you. This doesn't mean using a crowbar or steamrolling your way over others, but rather it means demonstrating the magnetic appeal of what you believe in and thus converting and corralling your audience. You do not require a lot of intellectual arguments to win a following, though a few classic and well-chosen catch phrases help immensely. It is not a debate but a demonstration that wins the day. If you believe in yourself strongly enough, that will be good enough. Without that, the best arguments are hollow.

Mercury Square Pluto

Avoid the temptation to squander inner resources, especially mental resources, this month, as it will be all too easy to get into time-wasting competitions that swing around closely held prejudices. Let others you know are wrong figure it out on their own time, and you do the same. Confrontations are at cross-purposes and don't make forward progress. Issues raised now will become dated all too soon, and who wants to fight about something that will shortly be irrelevant? Beware of overspending, both of money and of personal resources, as the compulsion to pursue beyond diminishing returns can sap your energy, deplete your pocketbook, and make you look foolish in hindsight.

Mercury Conjunct Ascendant

Rising Mercury means you lead with your mind, and whoever gets the first and best word in will get the most out of this period. A good line is worth its weight in gold, and a clever argument with an original twist will make you very much the center of attention. This applies not so much to writing, correspondence, and problem-solving, but more to what you say in person and how that sparks off the conversation and the thoughts of those around you. Don't speak without thinking first, but also have confidence that the first thought that comes to mind will be quite serviceable and will lead you where you want to go. It won't hurt to be glib.

Mercury Opposite Ascendant

Intellectual fireworks are most likely to be found in a partner, so be willing to play second fiddle to someone who is loyal and will carry you along and include you in the conversation. Use yourself as a foil for another's ideas, a mirror to their imagination to help them develop their own clear vision. Do this selflessly, and it will be as if it all came from you in the first place. Friendly sparring will serve to increase the intensity, but remember to give way when the stronger hand is not yours. It's a good time to learn what a two-way stretch is really all about and how depending on another brings your own message home. It's definitely a time to play doubles.

Mercury Sextile Ascendant

The way to good communication this month is less through making strong points and more through cooperation and the combining of strengths to come up with a successful consensus. The creative use of alliances is very much at the forefront, and when you can scratch each other's backs, everyone will be satisfied. Easy dialogues and conferences are very much the way to go, with lots of give and take, and secure in mutual goals and the ability for all to reach them in harmony. Friendliness is more important than getting everything right the first time, and others will iron out your mistakes (and vice versa) if you don't try to hide them but move with the common task.

Mercury Trine Ascendant

It may be easier than you know to say the right thing and have the right opinion right now, at least as far as others are concerned. Offhand remarks, extemporaneous comments, and relaxed observations all will hit the mark, though the last one to notice it may be you. So, when someone gives you a compliment, don't just toss it off with grace, but take a closer look at it and analyze its real value. You may be pleasantly surprised to discover an approach you can mine further. Remember some humility, however, as this may not repeat itself so easily, and it is partly the effortlessness of it all that gives it its value.

Mercury Square Ascendant

Look out for crossed wires and mixed messages for a while, as what you mean to say and what you appear to be saying may be confused or misconstrued. Be ready to spend some time clarifying your statements. In this case, forewarned can be forearmed, as you'll do much better to refrain from hastily commenting on anything you're not sure of. An empty mouth beats one with a foot in it. This may involve conflicts between home and work issues, between what serves you and what serves others. You may find you cannot serve two masters but have to pick which one is truly long-term. Serve both and you may keep neither.

Mercury Conjunct Midheaven

Pick your best and brightest ideas this month and spread them around, as your visibility is especially high. What you say, write, and think will be relatively long-range missiles, so if you think small you'll be wasting your powder. Spread the word about your accomplishments and you'll get returns in proportion. This applies less to new actions you take now or will take soon, but more to what you have already done that cannot be gainsaid. If that requires shortchanging your personal life in the short run in order to consolidate your position, so be it—strike while the iron is hot. What is heard about you now will have far-reaching effects on your career.

Mercury Opposite Midheaven

What you observe and learn on the home front may either interfere with or take precedence over your professional and outside life, and that may be just as well. When you forget where you came from, your external efforts cannot be sustained for long. When you renew your foundations, patch the cracks, and repour the cement, then you can sally forth with confidence, knowing your roots are solid, your support is firm, and you have refuge when you need it. Speak clearly to these issues now, and they won't pop up again later when you're too busy to tend to them. Know your heart, know your family, know your faith—then learn the rest.

Mercury Sextile Midheaven

Look for community support for your efforts, especially when it comes to praise for your past accomplishments. It won't help so much to promote it, just encourage it by complimenting others and sharing your expertise. What you give, comes back, and the less conspicuous you are about it, the better you will be received. A few good words here, a couple of favorable comments there, all mount up to the kind of recognition you would like to have without blowing your own trumpet. Internal success is between you and yourself alone, but external success is entirely in the eyes of your beholders, so court them with subtlety, treat them with respect, and you'll receive as you have given.

Mercury Trine Midheaven

People think better of you than you may realize, and much of what you have said without giving it a second thought may be treated as gospel by those less versed in what you do. This is a time to capitalize not so much on your latest fledgling efforts, but on what it is that you already do really well, what comes second nature to you that you can accomplish without a thought. Where you may have neglected old talents and successes, take time to capitalize on them and find new ways to apply them. Recycle, repurpose, rerelease, and resupply. Relying on older talents will both refresh them and give you a renewed resource you can depend on to support your next steps.

Mercury Square Midheaven

What you are saying and what is being said about you may be somewhat at cross-purposes right not, so untangling misunderstandings and setting things straight will be worthwhile pursuits. This particularly applies to conflicts between your career and business commitments and your close personal relationships, which could see each other as counterproductive at this time, leaving you between a rock and a hard place. If you are very nimble, you could have your way with both, but don't count on it or you could get tripped up. A good dose of honesty can go a long way towards mutual forgiveness and understanding all around, if you are willing to forgo ego trips.

VENUS ASPECTS

Venus Conjunct Mars

This is an apex time for love, romance, personal magnetism, and scintillating charm. Both desire and opportunity come together in the same basket, optimizing the opportunity for you to star in your own personal play, to slam-dunk the screen test of life and take your shot at the high life. At the very least, you will find personal and physical attraction heightened all around you, with everyone and everything taking on a greater aura of charisma. The world of the senses takes a front-row seat, with energetic amorousness center stage. Remember, though, that beauty is only skin deep, and this, too, will pass, so don't throw away something valuable for a moment's gratification.

Venus Opposite Mars

What you want and what you are able to get may occasionally be at odds, making the grass look greener on the other side of the fence. Just when the object of your desire is in your grasp, you run out of the energy or interest to pursue it. Or, when you're up and raring to go, there's no place worth going to. That can apply in almost any facet of life, but the feeling will be quite physical and sensory. Don't give up, however, as this is what rainchecks were made for. Make sure, wherever you go, that you're welcome back when circumstances change, and you will have used the time well setting up future rewards. Don't take it personally, but put your positive personal mark on it.

Venus Sextile Mars

This is a very good time to get what you want and want what you get without having to struggle too hard for either. Desire and willingness go hand in hand as long as you take a gentle approach and give partners plenty of room to be equal participants. Use a slow and easy hand, and you'll get plenty of strokes in return. It's a better time to show mutual appreciation through expected actions than through bold moves, so flowers or candy work better than something original and outrageous. Love through friendship is favored over passion through conflict. Love after a romantic dinner beats sex after a fight, which is not always the case, but is now. Don't push, just let it happen.

Venus Trine Mars

It will be easier than usual to let it be known what you want and have it returned in kind with love and even passion. Desire and opportunity complement each other, and whatever pleases is more than okay by everyone. You make it seem easy, even when at other times it is not, and that's part of pulling it off. The look and smell of success and ripe experience attracts more of the same, even if you've had none of either. Look and act good, and you are good. Let 'em know it was your first time some other time. Shoot first, ask questions later, and everyone will be the happier. It doesn't always work that way, but right now it does. Enjoy the bird in the hand; its heart is beating.

Venus Square Mars

Sometimes when you get what you want, you tire of it, or you tire of trying to get it. Desire and inclination are at cross-purposes, and it's easy to argue about the present and thus forget to enjoy it. Unless you are both watchful and tolerant, you may look back on this period as one of lost or squandered opportunities. There may not be much you can do, as half of the problems are not caused by you, but you can be a peacemaker and ameliorate them. Pull back from turn-offs, closely observe other people's needs, and put on a happy face even when you're disappointed in order to cut a downward spiral. You may not be able to get all you want, but you can and should cut your losses.

Venus Conjunct Jupiter

A big appetite—for things, for people, for life—characterizes this period. This can range from full and pleasant satisfaction to biting off more than you can chew and the resulting heartburn. So, satisfy your desires, which will be enlarged, but don't bust your gut doing it. The opportunities to increase your doses of love and money increase, and good feelings will generally abound—generosity, laughter, a party atmosphere. The best way to enjoy it is to savor each experience and not try to wolf it all down at once. Pick and enjoy the best of each course, but don't feel you've got to polish off every dish. Be a gourmet, not a gourmand. You can pig out on fast food later.

Venus Opposite Jupiter

Overdoing it or not getting enough of what you want can alternate, and striking a balance will be the challenge. A practical approach is what's needed and the last thing that comes naturally. The inclination to grab the brass ring or fall off the carousel trying is the first feeling to take hold, but after a few tries you realize you do better by reining in your desires and maximizing your possibilities. What or who you can't get now, put on layaway until you have enough, and then decide if that's really your best choice, anyway.

A positive, expansive approach pays dividends, though you should take care not to give more than you've got just because it feels good.

Venus Sextile Jupiter

This is a good time for developing resources and having a very enjoyable social time. Merriment and jollity can be pursued without fear of intemperance, because it's done in a social context where good company is more important than self-indulgence. Throughout, that's the theme—sharing the wealth, one for all and all for one, the rising tide floats all boats. It's really a time to learn the lesson that greater prosperity comes from creating win-win situations, not from looking out for number one. When one hand washes the other, everyone cleans up. Further, when you don't put the pedal to the metal, you get better mileage.

Venus Trine Jupiter

Whatever bounty you may have, this is the time to revel in it. The harvest of life is increased and appreciated right now as rewards are enjoyed and plans made to further increase and plenty in the future. Take more than a few moments to relax in good company, rejoice with the ones you love, and save dieting and belt-tightening for another day. There is an art to good living, so let your artistry show and share the benefits. The essence, however, is not physical wealth and possessions, but the ability to live life to its fullest, neither requiring nor asking for more than you have. When you can do that, five loaves and two fishes feed the multitudes, and the local natives show up with dessert.

Venus Square Jupiter

Wanting more than there is to go around can create an illusion of want this month, which can look like a thin camouflage for greed. The solution to a half-empty glass is to make it half-full, which requires nothing extra at all. There is also a hint of throwing good money after bad and not knowing when to stop. The cause is often not raw avarice, but failing to notice the cues of fulfillment, or missing the message that the party's over. In a phrase, don't top off your tank; it will just cause a stink. Hurt feelings abound when one person snatches up the last portion, so leave a little, if only to be polite.

Venus Conjunct Saturn

Doing more with less can be a theme now, offering lessons that will increase what you get later by learning to harbor your assets better. The essence of the situation is not usually real want, but spare resources compared to the needs of the job at hand. You could be short a nickel or a million, but the principle is the same: Make what you've got go further. In another sense, it means you can't always get what you want, so learn to make do and make it better, whether it's economic or personal needs that are concerned. This has the effect of intensifying desire through delayed gratification, so that when it finally arrives, you savor it all the more.

Venus Opposite Saturn

Untimely interruptions can put a damper on your party every now and then, but don't let them bring you down, just start up again after they are resolved. This can stem from an unfulfilled agreement or a failure of resources, like running out of gas because someone forgot to check the tank. Knowing that, make a point of having a backup ready or time to kill until help arrives. When it involves personal issues, avoid pointing a finger of blame, as hurt feelings or disappointments can and should be turned to opportunity by making a positive response. However, what you don't want, you don't have to keep, and this period can be a time when you sort that out and cut out the deadwood.

Venus Sextile Saturn

Extravagance leads to want, but moderation leads to plenty—that's the upbeat message here as you find that doling out your pleasures (or resources) makes you savor them more, especially when it's a shared event. By rejoicing in smaller things, you build to bigger ones and lay the foundations of friendship and trust that can last a lifetime. "Not yet" is just the preface to saying yes and should be enjoyed as the promise it truly is, not as a refusal. This period suggests courtship more than consummation, and foreplay in anticipation of continuing and increasing desire. You've got to set the stage before the play can commence, and a well-laid plot offers the most exciting climax.

Venus Trine Saturn

You have to know what you want before you can expect to get it, and this is a time for consolidating your needs, collecting your desires, and knowing what you have that you can really bank on. Personally, this means leaning on a lover or friend and enjoying every moment of support. Professionally, it means knowing what you realistically have to work with and can depend on to see you through. It can be a period of increase, not of wealth but of security, which is worth far more. When you have earned what you have, you take greater pride in ownership and are more likely to keep it. Now is a time to recognize those things, quietly rejoice in them, and then move on to more.

Venus Square Saturn

If you have to field some dissatisfactions this month, it is important to look at matters in context and not as black-and-white events that stand alone. Anticipation and expectation are as often the building blocks of disappointment as of fulfillment, and unrealistic desires distract from a clear view of what you really have. Therefore, you may be better off taking inventory of what is truly yours and dismissing fantasies that have not come true. Avoid getting caught in negative spirals when you find yourself at cross-purposes with your ambitions. Cut free of unwanted baggage and stay your personal course.

Venus Conjunct Uranus

The chances of surprise and adventure are high, especially if you are willing to give way to sudden changes of taste or desire, kick off your inhibitions, and try something new and thrilling. This can involve the unexpected appearance of someone new in your life or the opportunity to indulge in experiences that might have been beyond your previous norm. It can also mean an abrupt change of course in your financial resources—not necessarily up or down, just quite different. The way to make the best of all or any of this is to take an adaptable stance and enjoy the gusty weather. Putting up too much resistance can make you miss the boat, or capsize it.

Venus Opposite Uranus

If you were reserving the right to change your mind before, you can double that now. Expect sudden reversals of your desires and intentions that may surprise you and, particularly, others. Do things on the spur of the moment, then drop them a moment later in favor of something new. Can't make a decision? Flip a coin or let someone else roll the dice. There is a fine line between a life full of exciting new options and just a spate of erratic behavior, and you will have to be discerning. Originality and flair are valued character traits, but fickleness is a weakness. Try everything, but eventually settle on something.

Venus Sextile Uranus

Variety is the spice of life, but too many spices spoil the stew. Now is a time to demonstrate an even and sensitive touch that kicks life up a notch without overwhelming the ingredients. The key to this is integrating other people's tastes with yours and together coming up with a dish everyone likes. That applies across the board from your most intimate personal life to proceedings in the workplace. A balance is available now that will increase desire, enhance satisfaction, and introduce new and original elements into the mix. Controlled experimentation leads to perfection, and two or more heads are better than one. In this instance, only too few cooks could spoil the broth.

Venus Trine Uranus

You can experiment with a pretty free hand this month, as probably anything new and original you come up with will be pleasing to all concerned. By a casual display of talent, you can make the most far-out combinations seem tempting to swallow and the new pick of the day. Cutting-edge flavors become the rage all around, but you can be the one charting the course if you decide to right now. Whatever you want, just give it a try and don't feel you have to hold back. In fact, the essence of success is to give way to new forms of delight and then spread the wealth to the less creative.

Venus Square Uranus

Internal needs can be a little edgy this month, which can lead to unprecedented excitement in personal affairs that push the envelope. Knowing when to stop is the challenge, as the thrill of the new can quickly turn to overdose and disenchantment if good judgment is not maintained. You can court extremes, but don't marry them. Play with fire, but don't fall into the flame. You're not looking at disaster if you fail, just wearing out a good thing by satiation and its subsequent irritation. Pulling back from the edge in the nick of time is perfection, leaving you with a shiver down your spine. Teetering on the verge is where the rush is; running off the road will just land you in a ditch.

Venus Conjunct Neptune

Seeing the world through rose-colored glasses is the essence of romance, so that might be the direction in which to turn. Idealized beauty is what you'll both want and get, and it's not likely to be found in the accounting department (of course, who knows who might work there?). Although age and subsequent events take the bloom off the rose, it is the essence of the flower you seek and remember, so it's worth everything to throw yourself into as close to a mystic state as possible when the opportunity is right. Creativity and connection with spirit can flourish now, on whatever plane you choose or offers itself. Distill it, decant it, and delight in it for life.

Venus Opposite Neptune

What you were sure you wanted may not turn out to be what it was cracked up to be—or it may. The ground is a little shifty this month, and being sure of what you want may not be so easy. Seeming realities can dissolve into smoke and mirrors at the strangest times, so hedge your bets. Be prepared to change your mind or simply turn your back on an unsatisfactory arrangement. Promises are easy to make, but harder to keep, so indulge in dreams but don't marry yourself to them until they have had time to prove themselves. Conversely, if you aren't sure, don't fake it or you could fall into a trap of your own making. Honesty pays, even when you are not sure of the truth. All you have to do is say so.

Venus Sextile Neptune

In order to turn your imagination into reality, you need to share your dreams and make them apply to those who help fulfill them. This is a time when you can do just that; when you can stay on this side of the line between reality and illusion while living out mutual hopes and desires. This does not mean settling for second best, but it does mean sharing a second opinion. Actually, it is the checks and balances of more than one set of desires that keep both aflame but not burned out. This will apply to multiple facets of your life, not only the intimate part. The more you accept help in beautifying your world, the closer it will become to what you want and imagine.

Venus Trine Neptune

Separating out what you ideally want and what you practically can get is easier than usual for a while, and in fact it may seem to some that you've got it all without having to struggle at it. Be slow to shatter that illusion, as it may work for you more than you know. Further, it may come closer to the truth than you might have thought after you take a look at just how far you have come in beautifying and enjoying your life. There is much you may take for granted in your own world in the way of love and spiritual recognition that others would give their eyeteeth to enjoy. To ensure that the grass is really greener on your side of the fence, invite someone for a lawn party.

Venus Square Neptune

Chasing elusive butterflies can be a great temptation right now, but when you catch them, don't be surprised if they're just bugs. What you choose may not be what you want when you get it, either because you weren't seeing clearly or you have been sold a bill of goods. This is a very prudent time to look a gift horse in the mouth unless you're positive it's got a clean bill of health. "Perfectly clear" intentions right now may be murkier than you think. On the other hand, indulging small illusions in the interest of keeping the peace can be the humane thing to do, as long as you're sure your white lies have a silver lining. It's required.

Venus Conjunct Pluto

What or whom you've got to have, you've got to have, and you're likely to push for it this month. Desire is pumped up and passion will have its way, but try not to roll over anyone else in order to get what you want. A deeper understanding of what you want and must possess is in your grasp if you are willing to get below the surface and figure out the root causes for recurring needs. Control and self-control are two sides of the same issue that you might want to address, especially the gratifying aspects of personal power over others. Similarly, the power of possessions can be particularly strong, so you will have to remember the difference between owning and being owned.

Venus Opposite Pluto

A push-pull dynamic of desire can lead to excitement and push you into positions you might not have experienced. This could be at the romantic level, but it could also be connected to possessions you really want but can't make up your mind about. The essence is a tease—first a great intensity followed by a withdrawal, then back again, which heightens the encounter without bringing it to a conclusion. When you can't completely get what you want, you want it even more. You will have to decide if it was worth the effort to possess it after you've finally got it. Sometimes a bird in the bush is more appealing than a bird in the hand, and should be left there.

Venus Sextile Pluto

Steady pressure to achieve your desires will work well, particularly if you enlist the help of others. You can pull the strings from behind the scenes and no one will be the wiser, allowing you to piggyback on other people's more advantageous positions to achieve your ends. The trick is to be subtle and not steamroll your way in, but make others feel that giving you what you want is a total necessity for them as well as for you. In order to do that, you need to at least appear to share the wealth. You won't need to twist arms, just join hands and tug. You'll have to take the initiative, but once underway, things will take care of themselves without much further effort.

Venus Trine Pluto

You don't need to struggle or strong-arm your way to get what you really want this month, as it will come to you if you make it a matter of faith. Strong friendship and love will bring you what you need, so relax and enjoy. Spend your time promoting happiness, and it will be returned to you tenfold. Basically, it's the principle that if you don't appear to need it, it comes, while desperation drives people away. Trust in fortune, and fate delivers. Tempt fate, and fortune flees. Act like your heart's desire is already within your grasp, whether it be a person or a thing, and it will gravitate to you. Faith attracts, and doubt repels, so having faith is the solution right now, no doubt about it.

Venus Square Pluto

The temptation to pull out all the stops to get what you want runs strong, and it may seem that is the only way to go, but it isn't. In fact, the more you push, the more it eludes your grasp and in the process you may destroy the very thing (or relationship) you seek to gain. On the other hand, relationships that beckon powerfully may thrill for a time and overfill with intensity, but beware when the initial glow fades—there may be little there. Passion fueled by conflict and desperate desire is often the hottest, but fades suddenly and often regretfully. Friendship fueled by patience and love lasts a lifetime. You may be called upon soon to tell the difference.

Venus Conjunct Ascendant

The primary qualities to lead with this month are good looks and charm. You will be able to look more desirable, interesting, intriguing, and generally presentable, and that will be your wedge into every conversation, your foot into every social door. So, don't hesitate to spend a little extra emphasizing that, whether on personal care, clothing, accessories, or couture. Make the most of what you've got and improve on it as well, since it will pay you back in the long run. It's not vanity, it's keeping your tools in good repair. Further, by being the first to radiate love and affection, you'll find that coming back to you almost instantly. Whatever you're selling, do it in person for the best results.

Venus Opposite Ascendant

This month should be spent getting the most out of a partner or perhaps finding a particularly appealing new one. Whatever the case, you will find your bread buttered from the other side, and the person you pick will do well for you, indeed. Feel free to spend on your partner in the way of personal care, clothing, accessories, and so on, because the better he or she looks, the better off you will be. For the moment, this is the goose that lays the golden egg, so the best of care and feeding are certainly in order. In return, expect more than love; expect that your world will be expanded and your opportunities and contacts increased personally and professionally.

Venus Sextile Ascendant

Casual good looks are the way to go in personal presentation right now. You needn't come on like gangbusters or dress to the nines to make a good impression, but classic simplicity will give you just the right image to get what you want. The essence is not to stand out as something different, but rather to do what everyone else is doing, only better. In fact, if you ask for advice from others, you'll actually get something useful and they'll be complimented by the request. Further, they'll think the better of you, since you have taken a little of them and added it to your approach. Give and return compliments freely, and everybody glows.

Venus Trine Ascendant

There's probably not much you can do to make yourself look bad this month, so be as carefree in your personal presentation as you like. Your very lack of a fashion statement will be construed as a new look. In general, it's a comfortable time to relax and play whenever you get the chance, and you'll wear both love and pleasure with ease. You may also find it easy to chill and do nothing at all, but that would be a waste of this month's potential for enjoyment. Get up and treat yourself and someone else special, and you'll have a lot better time than you ever thought you would. Get up and get around, whether just down the street or to another country. You'll enjoy the surroundings.

Venus Square Ascendant

It may be difficult to decide whether to pay more attention to yourself or to your home and professional obligations. The latter will profit you in the long run, but you may have less fun in the short run. You can expect good things to be told about you from afar, but the same compliments may not be said to your face, so don't worry if you don't seem to be getting the response you're looking for. You'll look better down the line if you pay more attention to what is really important and put the search for immediate gratification on the back burner. Part of it may be that you aren't really sure what you want right now, but you'll know later.

Venus Conjunct Midheaven

This is a great month for getting the good word out about yourself, and the chances are you'll be receiving compliments for your professional achievements. Make sure you make the most of them to put you forward another step. Further, you can use your reputation as a wedge into personal affairs, where you'll have an edge over others who are less in the limelight. If you've been wanting something you couldn't get or find, now is the time to broadcast it all around, and, since your signal has a longer reach right now, someone will ring you up with just what or who you were looking for. Act like a risen star, because for this month at least you appear to be one.

Venus Opposite Midheaven

Your bread is buttered at home this month, and the more time you can spend there, the better. This can mean beautifying your surroundings, enjoying and reinforcing the closest relationships around you, and generally sprucing up the old castle. Chances are, that's an easy assignment, as it's what you'll want to be doing right about now. Nevertheless, it means putting some things on hold at work and perhaps earning a little less for the duration. In the end, however, it should pay off on both fronts. You'll know who loves you at home and just who misses you at work, and why. Both your emotional and professional bottom lines are likely to be the clearer for it.

Venus Sextile Midheaven

The best way to promote yourself this month is to get others to do it for you—rather, to help others do it with you, because it will have to be a cooperative venture. You may be surprised at the support you have that you never knew about because you never asked.

Spending, or even borrowing, money to reinforce career efforts will be worth it, as will seeking outside investment or backing, which may be more available now. It will be better to move in smaller steps, incrementally achieving your goals, than to go for the whole shebang at once. You will appear to be more cautious that way, which will actually bring you more of what you need, but in smaller packages.

Venus Trine Midheaven

You could probably just lay back and rest on your laurels this month, but that would be a waste, since you're in position to consolidate your gains and earn yourself even more respect. You may find that self-promotion is quite unnecessary and even inappropriate, as others will do the job for you. Your function should be to appear to be as good as anything said about you, yet hardly to have worked at it. Appear to take success for granted (like it were owed to you), but in fact treasure it and take all you can get without appearing greedy. When it's your turn to shine, there's no sense in hiding your light under a bushel—let your light shine.

Venus Square Midheaven

You're probably going to want to pursue your personal life and that of your partner, but you may find that doing so can get in the way of your professional efforts. You may have to field and defuse inaccurate or perhaps jealous rumors that come from pursuing your privacy instead of letting it all hang out. Or, conversely, you may let it all hang out a little too much and have to explain it. In the long run, no harm is done, but in the short run you may have to waste some time clarifying things. Once you set the record straight, you can move on to more profitable pursuits with the advantage that you have a higher profile and are someone to talk about.

MARS ASPECTS

Mars Conjunct Jupiter

This month has buoyant, palpable energy that can carry your ball all the way down the fairway to a hole in one. Like having the wind at your back, you can sweep on in a flurry of excitement into big, new projects that seem to be limitless in scope. Anything you begin this month—a project, a personal commitment, a purchase—has remarkable get-up-and-go with more than a dose of good luck by its side. If there is anything to be careful of, it is overshooting the mark because you don't realize your own strength. Remember also, when making plans, that you won't always have this unusual power supply, so don't overcommit yourself to keep this pace up in future months.

Mars Opposite Jupiter

A tendency to watch out for right now is erratic energy control, where you overshoot or undershoot the mark because you are not sure of your ammunition. Big plans may be afoot and there is much potential in strong, original, independent departures from the norm this month, but a steady aim is essential, because a miss is as good as a mile in all-or-nothing situations that arise. Extravagance comes easy, but it can be a threat to your security down the road, so only use the resources you need and try not to waste them. You can triumph with a single blow (the best way right now), but it must be well delivered, right on target, and totally focused.

Mars Sextile Jupiter

Strong, regular energy fuels original ideas and moves towards expansion, especially in cooperation with others. Tag-team endeavors where you can hand off the ball in a round-robin of responsibility allow everyone to contribute a little and wind up with a lot. You can place your bets on growth this month—not not in a runaway spiral, but steady and reliable. Projects begun in this climate of careful optimism will live long and prosper, as will children born now. Hard work seems easier as the fruits of your labor will come with greater regularity and certainty. Like a ship in the trade winds, you are blown steadily to your destination under clear skies.

Mars Trine Jupiter

This may be a good time to collect some of the interest on the work you have done over the years and roll it over into new investments or original projects. Anything you start now will have a good, forward-looking foundation and will easily remain abreast of the times and not become outdated. It's quite as easy, however, to just sit around and enjoy what you've got, but that would be a shame, considering what more you can do right now. It doesn't take much effort to get your increased assets working for you—that way, they'll stick around and not get frittered away. That goes for the personal side as well—be generous when you can afford to be. A helping hand is the best investment there is.

Mars Square Jupiter

Sometimes things get so tangled up that you just want to drop it all and leave. You may feel like that now and then, as well-meaning but slightly misdirected energy combined with barely off-base thinking leads to unexpected tailspins of unimagined dimensions. In many cases, it is better to just stop and start over rather than endlessly unravel the scrambled situation you've gotten into. Keep your sense of humor, however, and you'll have everybody laughing at themselves and ready to go again for a better try. This is a good time for working out the kinks, but not so good for launching new endeavors or making final commitments. The image that comes to mind is that of a dog shaking off water—have a towel handy.

Mars Conjunct Saturn

This is a time to be in it for the long haul, where getting things done requires patience, tenacity, endurance, and stamina. That applies as much to your personal life as to the rest of the world around you. You will not get things done quickly, but you will do them well and have accomplishments to be proud of. Expect to be tired, but also expect to get that second wind that comes from pressing on anyway like a long-distance runner. Precision operations are favored, as is restoration, renovation, refurbishing—anything that puts new life into the old. What you spend a lot of time perfecting will last, and hard work, however slow it goes, will pay off in the end.

Mars Opposite Saturn

Occasional moments of energy crisis, seeming paralysis of action, and empty gas tanks may turn up more often than you'd like this month. That can be because you overestimated your energy or underestimated the task at hand. At any rate, count on having to take some extra, unexpected rest periods and thus add some additional down time to your schedule to provide for them. You'll have less problem with repetitive tasks than with new ones, so it may be a good time to get ordinarily annoying busywork out of the way so your decks will be cleared when you get a second wind. Frustration with slow progress is understandable, but not inevitable, as the pace will pick up soon enough.

Mars Sextile Saturn

This month will run entirely according to plan—if, of course, you had a plan. Energy is best channeled through pre-existing routes rather than trying to break new ground, and if everything is in its place, all will go smoothly and require little or no monitoring. This is not an ideal time to fly solo, but rather to have assistance from others to make sure that things run on schedule and you're not the only one trying to police the situation. Once everything is in place, you can lock up and leave without worrying that something will go awry, but just make sure everything is really nailed down before you do. There is such a thing as pleasantly boring, and now you can experience it firsthand. Enjoy.

Mars Trine Saturn

This is a great period for switching everything onto automatic, laying back in an easy chair, and chatting with old friends—taking it easy in every sense of the word. You can expect home and work life to be pretty well under control without you having to exercise very much control, so you may have time to kill that would normally be routinely spent tying up loose ends and putting out brush fires. It's the perfect time for hanging around with your close friends and family, not too far from the action, as you don't really have to do much but you also don't want to fall asleep at the switch, just in case. You should pat yourself on the head for being so organized . . . at least for the moment.

Mars Square Saturn

This can be a bit of a burdensome time when seemingly unnecessary tasks turn up that aren't really yours to contend with, or shouldn't be. Things turn up unfinished that you were sure you were through with, and operations that never should have begun finally have to be shut down. It's a period of picking up pieces and putting things back into order so they'll run better next time. You may be more tired than usual, not from overexertion but from having your focus brought into disarray by unasked-for distractions that sap your energy. Don't get cross or feel sorry for yourself—these periods happen to everybody, and once past, you won't have to repeat it for a long while.

Mars Conjunct Uranus

This month contains the essence of a blitzkrieg or an atom bomb, and as such should be handled with care. The ability to strike suddenly, unexpectedly, and with overwhelming power can be tapped, but should be used sparingly and wisely. In the most mundane sense, it can represent lightning-quick decisions about people or property that could be harmful and cannot be taken back, whether big or small. On the other hand, well-placed moves that are made with precision can be effective beyond all imagination. So, imagine yourself this month with your finger on the trigger—and don't forget, the same goes for half of those around you. If you can do good, act unflinchingly—if not, keep it holstered.

Mars Opposite Uranus

It's a hard time to distinguish quick decisions from rash moves, brilliant strokes of action from hasty blunders. Usually, unfortunately, it's hindsight that sorts them out. The urge to act is strong, and insight into what to do comes in flashes, but the trick is to put wisdom somewhere into that equation. Like a dog on a leash, instincts held back for a moment turn into friendly tail-wagging, while an instant release outruns the urge for peace and you get a dogfight. So it goes this month. Channel sudden urges and inspirations and direct their actions and you'll leap ahead. Rush to judgment and there's no telling what kind of fracas may ensue. Take the edge off, and look before you leap.

Mars Sextile Uranus

The ability to mark a steady course that is inspired but not erratic is heightened by your willingness to receive input from others and share your vision and the way it is implemented. Inspiration and vision are hard to share, so this is no mean feat. However, the sudden and often erratic insight of one becomes the shared, multiple flashing lights of all, and what once was chaotic becomes a unified, forward-moving vessel of knowledge and wisdom. Some things are better done in company. So, to the extent you can be a team player, despite your eccentricities, do so for the moment. You may find that like stones in a jeweler's polishing barrel, knocking yourself up against others smoothes your rough edges and reveals a shining jewel.

Mars Trine Uranus

When everything you say is original, it can seem quite humdrum. That could be the way you feel, or at least appear to feel right now. The ability to make the extraordinary quite ordinary and thus acceptable to the more conservative is accentuated this month, so use it with a will and don't take it for granted. Trying out new approaches and making them work, whether at home with a partner or in the workplace, is something that can be done more easily, and you can take large steps that might have been worrisome or awkward at other times. Fortune favors the bold, especially those who don't look twice but just proceed. That can be a risky approach at some times, but this time isn't one of them.

Mars Square Uranus

Put some extra space on your calendar for picking up after yourself, as it will be easier than usual to trip and fall because you didn't look where you were going. This could be physical accidents, which are easy to prevent with a little extra forethought and care, but it's more likely to be emotional and social, where foot-in-mouth syndrome can be a real factor, and stepping on toes can be the latest dance craze. So, when in doubt at all, hold back and take a second look. What you haven't yet noticed, because you only barely glanced, can hurt you, or hurt someone else. Responsibility is the armor of love. Let your love live to fight on.

Mars Conjunct Neptune

If you want or need to pull the wool over people's eyes, this is the month to do it. Being a slippery fish comes with ease, and anything anybody does right now is a little hard to see correctly, so take advantage of the free camouflage. But be aware that this has its down side, in the form of difficulty taking actions with clarity and a sure foot. This is great for diplomacy of all sorts, and you can clothe your doings in any costume you desire with believability. Covert actions are a breeze, but taking a firm stand is not, so given the choice, put off final decisions. It's easy to waste energy for lack of an efficient way to channel it, so when in doubt, refrain from action until the fog clears.

Mars Opposite Neptune

Actions you take can be easily derailed now, so take extra precautions where you step, as puddles and potholes are hiding in unexpected places. Misunderstandings or misstatements can pull the rug out from under you when you put deed to word, so when you think everything is perfectly clear, think again. This can be frustrating, as the effect is intermittent. Therefore, don't push your agenda unless you have doublechecked what you intend to do and verified that you are on firm ground. Because you may be slowed down by the necessary extra precautions, put some extra time into your monthly schedule so you aren't pushed into making moves you haven't had time to research.

Mars Sextile Neptune

You may be able to raise excitement and expectations just by a little speculation, a "what if" proposal that can have dreams stirred up anew. This is a situation where you will have to interface your own hopes with those around you and together provide the strength and follow-through that turns concept into reality. Don't expect others to follow you unless you make them feel like it was really their idea all along—but in the end, the credit matters less than getting accomplished what you wanted in the first place. Hints and suggestions work better than sweeping statements and position-taking. Subtlety of touch is what makes your approach so appealing.

Mars Trine Neptune

More than you may think, your ability to make fantasy become reality is in natural gear this month. That doesn't mean you can hitch a ride to Never-Never Land, but it does mean you ought to take a second look at your options and see if you don't have some new possibilities you may have been taking for granted. Simply by asking "Why can't we?" may lead you to the place where you can, whether it's intimate dreams put off or idealistic propositions denied. All may sail for a while if you are willing to give an easy push as if it were the most normal thing in the world. This month, you provide the pixie dust.

Mars Square Neptune

This is a time when energy drains can come from all sorts of directions and not be too easily spotted. Sometimes it's premature moves, other times strokes come just too late, and the spring in the elastic step can't be counted on as you bounce down the road. You could slog your way through this if you must, but you can also choose a park bench with a good view and watch everyone else going through it. You may find that Margueritaville has just sprung up all around you, and getting cogent action underway is a full-fledged comedy. If you've got the time to spare, stick around for the circus; if not, strike out on your own and do a little reconsidering of your own course.

Mars Conjunct Pluto

It may seem that the only way to get things accomplished this month is with the iron fist in the velvet glove—but maybe without the glove. Certainly forcefulness is in the air and a powerful smack may sometimes be what is necessary to get someone's attention. Nevertheless, remember that to the extent you decide to push, you may get pushed back with equal power, and that could be at best a waste of energy all around, and at worst a serious defeat. So if you must pick a target to mow down, personally or professionally, make sure it's one that can reasonably put up with it and won't turn around and flatten you in return. Given your options, go with skill before force.

Mars Opposite Pluto

Are things putting up resistance? Back off, go around, and avoid confrontation when seemingly irresistible objects stand in your way. The urge to push your way through may be strong, but the ability to accomplish it may be inconsistent, leaving you with a job half-done and therefore not of much use. When energy is blocked, the absolute key to surviving is not to waste an assault thinking you will win. Winning, in fact, is not what it should be about, only getting on with things. So if your partner, boss, shopkeeper, or general person on the street says put up your dukes, find a different avenue to cruise for a while. The only fights to accept are the ones you're dead certain to win well ahead of time.

Mars Sextile Pluto

Gentle pushing can be just what you want to do to get your way, while appearing to be simply going along with someone else. The forces at large are fairly compelling and conducive to pushing things through by joint consensus, but not by a single ego having its way. So seek subtlety and even subterfuge when necessary and all will go well without ever a personal affront being taken. Large, hard decisions can be substituted with smaller yet compelling ones whose buildup forms a tide that brooks no resistance. Strength in numbers is clearly on the rise, but you may have to be the person to put those numbers together.

Mars Trine Pluto

You can get yourself across like Babe Ruth with a big swing this month, an easy swat that knocks it out of the park without a second thought. The trick, of course, is to avoid that second thought, which could spoil it. Plan to count on ease and even presumption of success when you make your moves, and you'll connect with the ball. Fidget with your bat, and you're out. The bottom line is the faith that you know you will connect, that this is your time, and that you are simply on a journey already described and now just being enjoyed. When you lose the edge of nervousness, all is free-swinging play, and that is what makes for a joyful, powerful thrust, whether in personal or professional adventures.

Mars Square Pluto

Don't bring a knife to a gunfight, or a rifle to a bombardment. The temptation to mismatch levels of power is great and can only end up with someone on the ground, so to speak. This is really a bad time for confrontation, because you don't know your strength or your opposition, and you're likely to waste energy and then be overcome when you're out. False and pointless statements can try to resolve forcibly what otherwise could have been done by diplomacy or avoidance. Aim is poor, and ammunition is great—what a wasteful mix. This is a time to seek more peaceful solutions, even though you think you've got 'em in your sights. You'll gain respect for it, and avoid a fall.

Mars Conjunct Ascendant

Mars rising indicates a period where intense energy and active personal intervention will be the way things get done around you. Thus, when you have the choice, make a live, in-person appearance to make your point, don't just write it or e-mail it. Do not shrink from being forceful, but do avoid force itself, which has a way of backfiring. This can also make you a bit accident-prone, the result of an intense and narrow focus that may cause you to forget to look both ways before crossing the street, literally and figuratively. Remember your strengths, but don't push your envelope too fiercely, or you'll have to spend twice the time picking up the pieces.

Mars Opposite Ascendant

Setting Mars puts undue emphasis on seeking out and recognizing support of your partner, whose strengths should be allowed to pick up the ball and carry the day. In fact, you may find some considerable opposition coming your way if you fail to do just that, so be willing to share and take turns. Once done, you are also in a responsible position to see that others don't push too far or carry more weight than they can handle on your behalf. That means being there to calm upsets, soothe raw nerves, and prevent accidents that can arise from having too narrow a focus combined with extra acceleration. Walk that extra mile, even if your feet are sore.

Mars Sextile Ascendant

Use your energies to combine with other people's, so that when everybody pulls together, the really big things can be accomplished. Your ability will be to insinuate your strength in a way that is not so obvious yet fills in just the right spot to set the tempo and guide the efforts. You don't need to huff and puff and strain yourself, just choose the right point and push lightly—everything will move along like magic. Timing, not force, is everything right now, and if you can remember how you pulled it off this month, you can save yourself a lot of wasted heaving and hauling. Brains, not brawn, are your biggest muscle.

Mars Trine Ascendant

Your ability to apply your energies to creative new endeavors is strenghthened, particularly if the efforts are physical and involve your personal touch to make it all come together. You don't need to push or strain—whatever comes first and most easily to your hand will be exactly what to go with. And further, your greatest effect will not be in the initial contact but in a graceful follow-through that allows timing to add a measure of strength you didn't know you had. If you have to struggle or repeat yourself, you're in the wrong game—bow out and try another where that natural rhythm works for you and not only gets the job done in an original fashion, it gains you admirers.

Mars Square Ascendant

Halting delivery or lack of clear follow-through can get you into a physical tangle, leaving you unnecessarily drained and frustrated. The more you think about what you are doing, the more you sabotage your own efforts. Your clearer choices would be to back off and let someone else do the physical work or personal appearance, or just try on something you care less about and therefore can be a little looser with. The essence of the difficulty is distraction—either from yourself or from conflicting inputs at home or work—and the answer may be to make yourself scarce and do what you need to do alone where you don't have to keep an eye out over your shoulder.

Mars Conjunct Midheaven

Battles, such as they may be, are most likely to be fought out in the open with everyone looking on this month. As long as it's friendly competition, it's the perfect chance to display your style and prowess, particularly in the career arena. Remember that the first in with the most is the likely victor, so don't flinch or pause or you'll be down a notch from the start. Under any circumstances, it's a great time to throw added energy into your career efforts and get instant feedback from a wide variety of sources. When you do, however, make sure you have all your ducks in a row, as a well-publicized mistake can make you a sitting target for the competition.

Mars Opposite Midheaven

This can be just the time to throw your energies into pulling your act together at home, even at the cost of losses in your career setting. Indeed, it may not be entirely of your own choosing, a case where you have to run to "put out fires" and smooth over acrimony that has arisen in your own backyard that can only hamper the rest of your operations. The answer will not lie in simply stamping out or smothering problems that arise, but rather in building new and appropriate settings where creative energies can be played out successfully before they become negative frustrations and begin acting out. Pent-up energies make better use than refuse.

Mars Sextile Midheaven

You probably don't need to work that hard to be liked and well-thought-of about now, but you need to have a fairly skillful hand at it. It's not what you do that does the trick, it's what you do to get others to speak on your behalf that shows the master touch. Do favors for free, but put your brand on them, so people remember where they came from. Don't use the hard sell, just make yourself available and interesting. Plant the seeds of desire so that they grow from within and when others praise you, it will be their own idea, or at least feel like it. A small nudge at the right spot can start an avalanche, so look for just the right place or person and then apply the least leverage.

Mars Trine Midheaven

This is a good time to seek out and accept help and cooperation in career projects and any area of endeavor that especially affects your reputation. By being magnanimous, you will seem all the more accomplished for it and will attract support and praise for your efforts. You could actually do nothing at all, drop everything, and take a vacation with no ill effects, but since you do have the opportunity to further yourself, do so if it is convenient. The essence of what you should and will do is self-confidence and a feeling that whatever you do, it's going to work, just like it always does. And if that's what you're putting out, that's exactly what will happen.

Mars Square Midheaven

Watch your step and watch the steps of those around you, lest some accident cause a tumble and you wind up the worse for wear for it. Not physically, necessarily, but you could certainly get the blame for missteps of your own or others that could easily have been avoided with a little extra care and caution. There is a fine line between trying too hard and not hard enough that may be difficult to put your finger on, especially when action is required, and the challenge is to maintain forward motion and a sometimes precarious balance at the same time. Cover your rear at all times.

JUPITER ASPECTS

Jupiter Conjunct Saturn

This broad socioeconomic indicator portrays a current climate of social and financial change in which new ways of approaching life come up to challenge more traditional beliefs and methods, marking a year or so of turmoil and uncertainty as it all sorts out. Just as so many individual dramas were played out against the backdrop of World War II (which began under this aspect), your own play finds itself on a stage possessed by strong forces and currents, which if well-taken lead on to fortune. These are what the Chinese called "interesting times," less predictable and more heroic than most.

Jupiter Opposite Saturn

This generational aspect (it happens only once every twenty years) marks a major social and economic struggle between the new and the old, the traditional and the radical, in which both sides spar and harden against each other. It is very much a time to choose one and leave the other behind in relation to the larger part you mean to play in the big picture. Prejudices increase, along with paranoia and a general fear of the unknown, leading many to cling to false idols and defend or advance them dangerously. Where you can find movement that leads both to what is right and is also profitable, there you will become elevated above the fray.

Jupiter Sextile Saturn

This year is what might be termed a time of profit-taking and cashing in on positions you have taken morally and financially and of establishing newer and more solid structures on which to base both your income and your moral life. It is a time to dispense with prejudice but not abandon caution, to reach out to others without selling out to them. This is the overall background on which your personal life is being played, so you will have to roll with it where necessary, though you may not now, or ever, be in the thick of it. If you are aware of the larger picture going on around you, however, you can only benefit from the trends in which you find yourself immersed.

Jupiter Trine Saturn

You are in a roughly year-long period in which the world at large is, in the deepest sense of the word, coasting. There is great faith that God is in his Heaven and all is right with the world, which is, as often as not, totally unfounded, a disillusionment often met the following year. But, for the moment, in the background all appears to be working smoothly both economically and morally. This makes for profitable enough short-term actions, as long as you can pull out quickly. The real personal bonanza may be had by searching out the coming shift that changes everything and putting all your financial and moral assets on it.

Jupiter Square Saturn

This one-year-in-ten period is full of shifts and changes, struggles and dissatisfactions, and confusion of purpose or profit in the world in general. If you didn't think ahead and see this coming, you could find yourself at the mercy of changing storms. Although this may not affect you that much, it's affecting the general populace, so expect shifting sands on a regular basis. This applies not just to economics but to disagreements about the very makeup of moral fiber, right and wrong, which side you are on, and so on. Don't go for an easy fix, as a good one probably won't be available at all this year. The wise are waiting, not wading in.

Jupiter Conjunct Uranus

There is a fourteen-year rhythm of discovery that is now at its peak for the world in general, a time when big new ideas blossom forth into existence and can be yours for the picking. It's generally a time when risk-taking pays off if you really believe in what you're into, personally or financially. It is a real test of your talents, however, because everybody's in the same boat, rising on the same tide. It's what you see in it and how you use it right now that counts. It's a good time for starting new projects, bursting into new personal affairs, or exploding onto a scene you never knew existed a year ago. Just remember that time passes quickly, so lay down roots fast, and make a strong foundation with dispatch.

Jupiter Opposite Uranus

This is a time when large-scale efforts and the best of intentions can suddenly change, go awry, be shot down, or get reborn—not just for you, but for everyone around, so keep that in mind before putting all your eggs in one basket. It's a time of discovery not so much from new revelations but from uncovering what was wrong, popping the over-inflated balloon, spotting the phony. If it looks too good to be true, it probably is, and the profit goes to the one who finds that out first. The same can apply to your heart, where promising more than can be delivered is a distinct possibility, and its later revelation comes as a shock. Use normal precautions and get confirmation before you go whole hog.

Jupiter Sextile Uranus

It's a good time to invest in mutual discovery, where what you stumble on somehow perfectly dovetails with what just fell into your partner's lap. You don't need to push or pursue the far-out to go way into yourselves, and you'll probably even have a little help from your friends, because right about now they're probably doing it, too. This can apply to the boardroom or the bedroom equally, as subtle inspiration and small "Aha!" revelations snowball to give everyone a totally new view on an old subject and reinvigorate flagging energies and interests. The key to success: Be aboveboard, and don't go overboard. Share what's given; expect the same.

Jupiter Trine Uranus

This is, in general, a period when it's more acceptable to experiment with the unusual, pursue original approaches, and follow those flashes of insight, however strange they might seem. Partially, it's because what you might have thought strange not long ago now seems obvious and everyone wants a piece of the action. Now is not the time to recriminate or remind others of how unimaginative they recently were—much better to acknowledge how ahead of the pack you were because, of course, you're a natural leader. Only imply that, of course. In the flush of acceptance, do not forget hard-fought recognition that may seem easy now but may yet have to come again.

Jupiter Square Uranus

Attempts to leapfrog from idea to execution may be at sixes and sevens for a while, but it's not your problem alone, it's in the air. Troubleshooting and debugging are necessary parts of life's progress, and, like everything else, it all happens at once. This makes this not such a good time for leaping into new worlds when the old one still hasn't gotten its act together. Until you're clear about now and know you're on firmer ground, seek out the glitches, the overlooked pieces of the puzzle, the stuff that glues it all together into one working whole, whether that be a marriage or a merger.

Jupiter Conjunct Neptune

There is a feeling all around that the next step along the way is both a gift from the gods and sanctioned by them as well. That may be so, but those of us on Earth will have to live up to the ideals we set now, which may not be an easy or practical task. It can be too easy to expect too much of yourself and those around you because you don't include the rough but real aspects of reality and temptation we all have to live with. This is a great time to be a high priest—or to act like one—but make sure that what you sacrifice on your altar is not what you see in the mirror. The light plays tricks that way.

Jupiter Opposite Neptune

One of the greatest challenges of breaking new ground is establishing a clear vision and holding on to it, despite the vagaries of the mundane world. That's the challenge all around right now, as it is difficult to focus on what seem to be obvious improvements in life, love, business—you name it. Just when you think you've got it all together, the mind wanders, and the objective eludes the eye. If this seems likes it's contagious, it is, as it's a general trend of which you are only a part. To that extent, the most likely places to find what you are looking for are the quests of others who also are trying to find their hearts. The more you can share the search, the more likely each is to find the journey's end in sight.

Jupiter Sextile Neptune

A more realistic and satisfying way to blend dreams, hopes, and expectations by sharing is in the wind, if you are willing to take a deep breath. Actually, a series of shallow breaths describes it better. What once you may have kept to yourself, it is time to exchange with others who have been in the same situation. Fortunately, the air is very conducive to doing just that, and what might have been hard to say before comes more easily with the knowledge that it is a two-way street of mutual trust and reward. You may find this critical to your happiness, or you may just find yourself in a general situation where it helps others and it's the best thing to do. Either way, you win.

Jupiter Trine Neptune

If what you believe is what you are, then everyone is feeling very self-aware right now, and rather relaxed about it. There is a general feeling, in the background, that dreams and expectations are quite realizable (if not, indeed, already realized) and that none of it requires much tending, as what will be, will be; I'm okay, you're okay; win-win; let it be. That can be particularly annoying if it isn't applying specifically to you right now, but understand it as a background illusion that can be used if you're clever, or otherwise taken with a grain of salt, as the best of times change with time itself and all illusions devolve sooner than expected into the next scene.

Jupiter Square Neptune

This is a good time for taking that second look at what life seemed to offer, or that it claimed to promise. That doesn't mean applying to your life, specifically, but you'll probably find that life's illusions are the current question in the air and there is good conversation in it at the least. It may mean for the moment that optimism fails, hopes lead to disappointment, and a lot of other confusions result which are caused by both unclear desires and fuzzy thinking. If you can be on the outside looking in on this one, you'll be the happier for it. Unexamined expectations goeth before the fall, and where you can be a cushion to others and a shoulder to lean on, you'll be the one who benefits from it later.

Jupiter Conjunct Pluto

Sometimes innovation and expansion arrive on a gentle breeze, and sometimes on a hurricane. This time is likely to see the latter, when progress is not to be resisted or it will overcome you by sheer force. This does not mean you have to get on board every new brainstorm, but if you see a juggernaut coming, either climb on or get out of the way. On a personal level, however, this can mean shoving your ideas or beliefs down someone's throat, even though you mean well, and that can be tantamount to coercion or worse. No means no, whether you believe it or not, and since this is a tendency in the air in general, you may be the one on the receiving end to most benefit from that.

Jupiter Opposite Pluto

Sometimes the most perfect, wonderful, ideal schemes go racing along and suddenly hit a brick wall. Watch out for that about now. The possibilities of not-quite-irresistible-enough forces encountering truly immovable objects are rife. So, if a big "No!" suddenly stops you (or a situation you are involved in) in your tracks, it's probably not the right time to stand and argue and beat on the wall. That's because, right or wrong, you probably don't know the root cause of the objection and therefore haven't the key to getting around it. The best policy is to back off, without blame, and move on to modify your plan and execute it on more fertile ground. Don't throw good money after bad, a true heart after a false one.

Jupiter Sextile Pluto

This is an excellent time for whipping up great faith in whatever new trip you are on, regardless of its apparent merits. Faith is the operant word here, for if you instill enough of it in yourself and inspire enough of it in others, you can succeed where you probably wouldn't or shouldn't otherwise. Basically, your personal or financial success (it can be either) can hang on your ability to generate a deep feeling of trust and the inevitability of what you believe to be the right course. If you can make your belief theirs, then you can take a ride together and you won't have to sell anymore. And when you pull together with an abiding knowledge of success, it's a lot more likely to happen that way.

Jupiter Trine Pluto

New frontiers, hot prospects, and expanded operations are not hard to sell right now—in fact, the soft sell will be the most successful. Whether you are promoting your affections or a consumer product, act as if it's the only way to go in the world and is pretty much an accomplished fact already. The more faith you have in yourself, the more others will have in you, and the more likely things will work out right for you, since everyone is expecting them to. Still, keep on your toes, because if you believe too strongly in what you are saying, you may just sit around and wait for it to happen, when in fact it is you who must be the driving, though not domineering, force.

Jupiter Square Pluto

In your enthusiasm for a new cause or idea, you can sometimes back it up with a little too much or ill-placed force and break the very thing that you want to sustain. Or, you can overestimate the effectiveness and reach of your powers and wind up going a bridge too far. There is a general climate right now for making that judgment error, and once you have made it, there usually is no going back. You may equally find yourself the recipient of this kind of pressure, and should you see it happening, don't fight it and have it go to its logical hurtful conclusion, but don't give in, either. Just make yourself scarce until the situation blows over. In the survival game, the fittest often know to flee.

Jupiter Conjunct Ascendant

Jupiter rising this month means you can be the firstest with the mostest, and the larger the scheme you are promoting, the better. The more expansive you can be, the more psychic space you can take up in a room, the more successful you will be. That doesn't mean you should hog the spotlight; just do what it takes to make people want to train it on you—be jovial, upbeat, positive, reassuring, humorous, supportive, encouraging, and philosophical. If there is one potential drawback here, it is the urge to overdo, particularly to overeat, overdrink, or overestimate your physical prowess. Go to your limits, but don't push the envelope—and when in doubt, err on the side of caution.

Jupiter Opposite Ascendant

Setting Jupiter puts the ball very much in your partner's court, or certainly look for another person to be a major font of new ideas, motivation, and extra stimulation. This may be just the time when you have run short on steam or your creativity has gone a little stale and you could use some pep and inspiration. Give others the welcome space to provide that, and that is just what they will do, although you need to make it clear the door is open. When needs go unspoken, the needy often remain that way, when there is plenty to go around. So, don't stand on ceremony; fling open your doors and let someone else's sun shine in, along with a little oxygen and fresh air to give you a needed boost.

Jupiter Sextile Ascendant

You may find that your assets are a little bit greater this month than you had previously supposed. This could be because of an unexpected gift or investment, but it could also increase your necessity to spend in the process. The challenge will be for you to use your ability to sell yourself to pull others into your pot and help support the general treasury. You will be looking rather well, anyway, but all that you can do to make your personal appearance perfect will pay off double. Convince others that you are discovering a whole new thing together, and that it's someplace safe to go, and then you'll have backers aplenty. The touch throughout should be of charm and reassurance, not pressure.

Jupiter Trine Ascendant

Largesse comes easily this month, and the best way to see that you get plenty is to be easy and generous without asking for returns. When you ooze luck and confidence like it's your birthright, it will be, as looking good leads to being well-off. Incredible tales told with total self-assurance can turn into self-fulfilling prophecies, so if you stay upbeat, you'll be creating your own future. Visualizing what you want is half of it, but actually stepping in the right direction with a strong will is essential as well, so don't just expect it to drop into your lap—although you should act that way. Be modest, aim high, and behave as though you've already won.

Jupiter Square Ascendant

Beware the possibility of overstating your case or pushing your brilliant idea past the point of diminishing returns. It may seem that one more push ought to do it, when in fact one less is what's needed, and overselling yourself is as disadvantageous as underselling. This will be more the case when you are communicating in person than at a distance, so you may want to keep some space between you and the world, and let e-mail and snail-mail (or simply another representative) do the talking. If you can do your dealings from home while taking care of extra business there, you can be doubly blessed. At any rate, avoid exaggeration, and put extra humor into what you do.

Jupiter Conjunct Midheaven

Jupiter high aloft makes this a great time for expanding your career position and making the world aware of how well you're doing it. Your good side is going to hit people first, and they're likely to overlook or forgive your faults, so make the best possible mileage out of it. Get your foot in the door of new establishments, even though you may not want to go there. It pays to be known. Taking an in-person approach may not be the ideal or most efficient way to do this, so use all the outside help and media tools available to work for you instead. That's why they're there, and that's part of this month's lesson. Spread the news, and lose the blues.

Jupiter Opposite Midheaven

You are likely to find that there is some considerable room for improvement at home, and that's probably going to be where you get the greatest satisfaction, taking care of your castle and refilling your inner gas tank. This cannot help but affect what time you can spend pursuing career matters, which will have to be handled on a more sporadic and part-time basis, depending on how much time and concentration you can spare. The result may be that you have to push back finishing dates on some projects and add a little extra open time to your schedule to carry the double, alternating focus of the month. Inner refueling now, however, will maximize your mileage down the road.

Jupiter Sextile Midheaven

Look to outside support from higher-ups or financing to help put you on the map this month. The emphasis should be less on selling yourself and more on promoting what you are and what you do as something that raises the level of the general good. Remember, when people vote for you (or invest in you, personally or financially), they are ultimately voting for themselves and their own welfare. So, your thrust should be a common purpose, shared ideals and goals, and interdependent support. By presenting others with new opportunities, you create opportunity for yourself. Don't look at being open-handed as giving away the store. Think of it as free samples and promises of what's inside.

Jupiter Trine Midheaven

The more you seem to take your career successes (real or imagined) for granted this month, the more people will wonder just how you managed to achieve it all and will become unpaid advertisements for you. Emphasize your creativity, your ability to spend wisely, and just the joy of being you, but never in a self-promotional way. That is the best way of being self-promotional right now—why do for yourself what others can do (and possibly better) for you? That doesn't mean you should rest on your laurels, just don't flaunt them. In fact, now is a time when you can best feel out what people really think is good about you so you'll know better where your strengths lie.

Jupiter Square Midheaven

People may have a hard time this month differentiating who you are in person with what they have heard about you. No matter how you dice it, this is tricky to resolve without either seeming too self-important or making the situation even muddier. Two solutions seem best. First, you can avoid personal appearances and let your reputation and partners do the talking, or you can be extremely modest but warm and supportive in your interactions so as to minimize any misperceived threat (Think you're not threatening? Anybody can be, without meaning to be). Either can work, or even a combination of the two, but if you don't get the right response, don't take the same approach a second time.

SATURN ASPECTS

Saturn Conjunct Uranus

Saturn conjoins Uranus for a year or so once every forty-five years, providing a background influence that is particularly conducive to major steps forward in science and technology. The first atomic pile and the introduction of the Internet to the general populace are prime examples. It usually indicates introductions that explode (literally or figuratively) a couple of years later and have a profound effect on social and economic trends. The trick is, of course, to be in on the discovery or know what it is so you can shift your life accordingly. If you're not, you'll be caught up in it soon enough and most of the world will be your company.

Saturn Opposite Uranus

This rare (once in forty-five years) aspect provides a background of a year or so when new advances in theoretical science occur that take a long time to prove, thanks to an often intransigent scientific establishment. Relativity theory and chaos theory are two good examples, which took years to become thoroughly accepted and a part of the general consciousness. If you happen to know what they are, it probably profits you little, except you can tell your children you were there first, when they finally get hip to it. Nevertheless, it's worth looking around for, as the foundations for another wave of the future are being laid at this very time, which is no small thing.

Saturn Sextile Uranus

This is generally a period of a year-plus when there is a general background effect that helps commercialize new trends in science and technology or milk the last out of old ones—as opposed to introducing revolutionary new ideas. As such, it means the opportunity is out there for everyone to get a piece of the pie before the new stuff hits or the latest rage wears out. Whether that will personally impact your life in a specific way is a moot point, as everyone including yourself will be affected by this underpinning of commerce where the often contradictory sides of originality and conservatism are at peace for the moment in the interest of making money.

Saturn Trine Uranus

This year-long period, which happens twice every forty-five years, makes for a curious backdrop that everyone has to deal with but that is not necessarily individually specific. It's a time when radical and conservative trends, whether in politics or technology, are at odds but are strangely comfortable bedfellows, each relying on the other for fuel. It makes for a period of surface tension that belies the fact that not much that is new or solid is really happening—lots of smoke and mirrors, but surprisingly little substance. At the same time, a somewhat insubstantive peace reigns that has in it the seeds of real confrontation and revolution yet to come.

Saturn Square Uranus

This is a roughly year-long period that occurs every twenty-two and a half years that, although it may not impact you specifically more than others, is usually a pretty risky time to live in. It marks conflicts across the board between radical and conservative ideologies that tend to spiral and have difficulty finding peaceful resolution. Naturally, this produces an underlying tension in everybody's life that makes harsh realities harsher and good times more desperate. If there is anything to be learned from it, it is that compromise, not confrontation, is the only way to go. Every twenty-two and a half years, an alarming number of people fail to understand that.

Saturn Conjunct Neptune

This roughly year-long influence happens every thirty-six years to everybody, but some cultures experience it more strongly than others, and so do you if you are among such a culture. Being a confluence of fairly inimical principles, it outwardly manifests itself as decay and collapse, along with fertile possibilities for a new order to eventually manifest, since it is also a triple Jupiter cycle. The collapse of the Soviet Union is a case in point. On a higher plane, it can be a time when spiritual aspirations touch ground or, conversely, hit the ground. It's not something to hang your hat on, but it can teach you that the loftiest ideals must produce or perish, while from something rotten, the best crops grow.

Saturn Opposite Neptune

This approximately year-long period, which everybody shares once every thirty-six years, puts a backdrop of considerable deception and collusion in government circles followed by exposure and consequences. The Watergate conspiracy and the covert rise of the Nazi military machine come to mind. The fact that people can be so false and so blind at the same time naturally affects all and suffuses cultural morality and trust at a fundamental level. On a personal level, it means cover your rear and don't rely on what you hear, especially from the most normally reliable sources. What you don't know probably will hurt you, so check twice, and then again. Everybody's in the same boat.

Saturn Sextile Neptune

This is a roughly year-long stretch when life for society at large can seem steadier, when what is believed true can appear reliable and the powers that be for the moment either tolerate or coopt the spiritual world. Thus, it is sometimes a period of seeming spiritual progress. It seems like the bloom is on the rose, but it's really a work in progress that results from hopeful alliances and temporary conveniences. Although this is a universal experience and has little application on any one specific level on a month-to-month basis, it does raise the universal opportunity to make short-term gains out of what others may believe are long-term trends.

Saturn Trine Neptune

For about a year, the world in general seems to be in the process of merging spiritual aspirations with material possibilities. It is as if Church and State finally realized they both agree at heart. It is a period you will likely look back on with fondness and a sigh, realizing that hope does indeed spring eternal and eventually may triumph. But the meek aren't about to inherit the Earth, not yet anyway, so perhaps the best way to capitalize on this general trend is to associate yourself with hope, both yours and other people's, and where possible let it float your boat for the duration. Inner and outer stability flourishing together can only last for a time, like Camelot.

Saturn Square Neptune

This period of about a year, which recurs every eighteen years, is often a time of great hope or disillusionment for society at large, when what seemed to be a harmonious, hopeful, and supportive world turns back into a surprise Darwinian scenario, or vice versa. The period of the Kennedy assassination and the end of World War II are examples, and though not everyone is personally affected, everyone is somehow changed and life isn't quite the same afterward. The conflict of ideal with real, whichever temporarily triumphs, reminds us that these two worlds are fundamentally at odds and that we must juggle both in order to be human. The lesson lies not in the fact that we see it, but that we so quickly forget it.

Saturn Conjunct Pluto

This roughly year-long stretch marks a time when conservative ideas and organizations gain their power through force, fear, or covert activities, or by simply steamrolling their opposition. The McCarthy era is a good example of this feeling. If you're not specifically in the way of this sort of operation, then this probably will have little personal effect, but it puts an air of worry in the social background that things are not as stable as they should be and that something somehow is amiss. That even filters down to children born in this period, who are noted for anxiety and depression and the psychologically driven creativity that is sometimes induced in later adulthood.

Saturn Opposite Pluto

This roughly year-long stretch marks a time of populist opposition to the establishment by forcing change or preventing repression. It's a general society-wide phenomenon, often happening at different levels, over different issues, in different places, but it creates a mood and often a counterculture that greatly influences social development. If you're in the thick of it, it could affect you directly, but if not (and most people aren't, usually), it still colors these times and the feelings and opportunities that surround you in your environment. At best, it's an important and thoughtful growth period, and at worst, just interesting and entertaining times that you are bound to remember.

Saturn Sextile Pluto

This roughly year-long period describes a background in which previous social conflict has been resolved and the establishment and powers that be seem to be pretty much doing their jobs. While it may not necessarily make a personal impact, this rolling along of the status quo is something you may be able to use or at least be aware of to your advantage. It means you'll get more mileage out of being a cooperator and someone who does not rock the boat. There is more headway to be made running with the tide than against it, although you may not agree with everything you have to put up with in the process. The time for major departures from the norm is not now, but will come soon enough.

Saturn Trine Pluto

For a year or so there is a time when the social backdrop seems relatively peaceful, and it could be that the right people are in charge for once and getting what's needed done. Though this influence may not be specific to your life, you can make your life specific to it by riding the wave and taking advantage of calm waters to do some exploration you might not have done otherwise. This can include alliances and even friendships with people who differ from you considerably but are now more open to and comfortable with your way of thinking. It's sort of a "pax Romana," in which the successful control of the political and cultural matrix allows for commerce where there was conflict before.

Saturn Square Pluto

This period of a year or so marks a somewhat troubled time for society at large, during which political and economic struggles arise and engage in conflicts that don't actually seem to get anywhere but take up a lot of news time. Unless you're part of the news, chances are this won't hurt you directly, but it does sow a certain unease all around that makes people uncertain and less willing than usual to commit to long-range agreements. When the wind is shifting, everyone waits until it's changed before settling down to business. With that in mind, you may find that you can make more headway in smaller affairs and short-term commitments, personal or financial, pending further advisories.

Saturn Conjunct Ascendant

With Saturn rising you will probably be taking everything rather seriously this month, either because there are a lot of personal pressures to deal with or because there are important, long-term relationships that hang in the balance and depend on wise decisions on your part. Gravitas is definitely the watchword, and heavy hitting will be what's required to get yourself across. On the other hand, there are times when you could, and should, lighten up a little and chill the grave demeanor. You don't have to look so serious to be serious. It's the acts you do, not what you act like, that will see you through the situation, and too much posturing is only a distraction.

Saturn Opposite Ascendant

You may find your partner putting up resistance this month, either because of disagreements or an inability to get a move on. Patience is therefore required. However, it is a good time to make long-range partnership commitments, whether in love or career, by deciding who's in and who's out in your life. Clearing out the dross will lighten your load, but you'll only do it when you find you're trying to handle too much. You can look to the older and more experienced right now to help you decide, so don't be afraid or too proud to ask. There are times when you might feel less encumbered if you were alone, and clearing some space to be by yourself could help you see better what's really happening.

Saturn Sextile Ascendant

This month may well be a time when you have to rely on less support from those you expect it from, when you will have to lean on yourself and dig deeper into your own resources in order to cope. This can mean cutting back on expenses and deciding just what you really need and don't need. This does not suggest fighting off deprivation so much as learning to use what you've already got more efficiently and eliminating extraneous or redundant expenditures and routines. If you don't use it, lose it. Repack your life's luggage so it takes up less space and you can find things faster. Make your drill tighter, your response time better, and your firepower more accurate.

Saturn Trine Ascendant

If you can depict yourself as having already arrived, you will find you can get credit for what you have done without having to significantly add to it right now. You can get a lot of support if you just stay put without positing yourself in a more extended position than you occupy right now. Look to consolidate your personal gains by going easy and holding back even if tempted to say more. Do the same where cash is concerned by avoiding further credit debt and staying away from speculative investments. Look more experienced, but not older; more serious, but not dour; more conservative but not fossilized—someone who is above the fray and doesn't shoot from the hip.

Saturn Square Ascendant

Be careful that you are not tripped up by a lack of self-confidence this month, so do what you know well and skip the live experiments. Repeated attempts to improve the situation will probably fail, so if you don't get it right the first time, don't beat it to death, fix it later. You may feel slighted in an incident at home or because of something someone says about you, but don't give it much weight, as you're probably working off wrong information that will be cleared up later. If you don't get the attention you want, be patient and don't take it personally. Similarly, avoid unduly harsh criticism of others right now as you will only end up with egg on your face if you turn out to be wrong.

Saturn Conjunct Midheaven

Saturn overhead this month could get you into trouble through ill rumors and resistance in career matters. The way to avoid these is to head them off at the pass using surprise tactics. Be the first one to uncover your own errors, and you'll be said to be mature, not just mistaken. Claim responsibility for the past and then move on. When challenged, do not confront, but draw in your adversary with a quick retreat and then move on past. Intransigence will get you nowhere, so don't feel you have to prove a point. Keep your eye on what you really want in the end, not just right now. When it's clear you're in it for the long haul and have the necessary tenacity, criticism turns to admiration.

Saturn Opposite Midheaven

If you spend too much time tooting your own horn, your harshest critics will be right where you least expect them—at home where everybody knows who you really are. So, take advantage of the trend to learn where some of your weaknesses may be and where you need to get back to basics and return to your roots. Don't be afraid to seem old-fashioned in defending what you believe—oldest is often the best, if it's stood the test of time. You may also find that you have a way to go before you make your mark, as you begin to see that quality is worth waiting and working for. By overcoming obstacles one at a time, you build a well-paved road to the future.

Saturn Sextile Midheaven

Look for surprise support for your ventures from behind the scenes, from hidden resources and places that you had long ago forgotten about—not anything sudden or overwhelming, just the steady push of those who may not want to wave your flag but are happy to see it up there. Let this deep background swell remain largely out of sight, and you'll continue to work with it. Blow its cover, and it's goodbye. Silent partners, anonymous donors, and discreet investors can all give you the extra capital, financially or professionally, that you need to do the job really well. Be known as a promoter of quality without gratuitous innovation, and you'll keep those backers on your side.

Saturn Trine Midheaven

Hard work pays dividends—or at least that's the image you'll get this month with only a little effort. The image to portray here is of one who spends less and has higher productivity as a result of filling all the unused spaces in your schedule. Similarly, you'll find you make more and get more cooperation from compatriots in the workplace if you pursue this image. This doesn't mean turning over a new leaf, just showing off the benefits of how you do things already. Don't expect to cash in overnight on your reputation, but be sure that as you are known for your steadiness you will be rewarded accordingly. The point is not to bust your chops, just pick up every stitch.

Saturn Square Midheaven

Difficulties with personal relationships can put a dent in your ability to pull off career moves and can generate less than complimentary fallout unless you go out of your way not to seem too critical or put up undue opposition to other people's needs and demands. This may not be entirely within your grasp, as it may have to do with someone else's failings that you cannot do much about. Expect to lose some ground, so make sure you have the ground to spare in front. What you can afford to lose is simply a gift to Fortune, which Fortune repays in her own good time. In the meantime, be patient with yourself and with others, as misunderstandings will repair themselves if you don't press them.

URANUS ASPECTS

Uranus Conjunct Neptune

This three-year period (most recently from 1992 to 1995) occurs only once every 172 years and is reputed to have the effect of a spiritual awakening upon society at large. Too lengthy to have much differing effect monthly, it nevertheless sets the tone of the period. Uranus/Neptune's relationship to the rest of the chart by house and aspect give it a united, two-for-one punch that is both jangling and soothing, wakening and soporific, like trying not to sleep at the wheel on lots of caffeine. So, look to its relationships this month to the rest of the elements in the Lunar Return to see how this really "big picture" planetary aspect works for you.

Uranus Opposite Neptune

This three-year period (most recently centering around 1907) occurs only once every 172 years and is reputed to have the effect of spiritual and scientific tumult upon society at large. The last time, for instance, it just followed on the introduction of Einstein's first theory of relativity. Too lengthy to have much differing effect monthly, it nevertheless sets the tone of the period. Uranus/Neptune's relationship to the rest of the chart by house and aspect give it a back-and-forth punch that is both jangling and soothing, wakening and soporific, like trying not to sleep at the wheel on lots of caffeine. So, look to this aspect's relationships this month to the rest of the elements in the Lunar Return to see how this really "big picture" planetary aspect works for you.

Uranus Sextile Neptune

This several-year period (most recently centering around 1966) occurs only twice every 172 years and is reputed to have the effect of spiritual progress upon society at large. Too lengthy to have much differing effect monthly, it nevertheless sets the tone of the period. Uranus/Neptune's relationship to the rest of the chart by house and aspect give it a reinforcing effect that is both inspiring and soothing, wakening and tranquilizing, like a rum and Coke. So, look to its relationships this month to the rest of the elements in the Lunar Return to see how this really "big picture" planetary aspect works for you.

Uranus Trine Neptune

This several-year period (most recently centering around 1939) occurs only twice every 172 years and is reputed to have the effect of spiritual and scientific progress (though not without strife) upon society at large. Too lengthy to have much differing effect monthly, it nevertheless sets the tone of the period. Uranus/Neptune's relationship to the rest of the chart by house and aspect give it a reinforcing effect that is both inspiring and soothing, wakening and tranquilizing. So, look to its relationships this month to the rest of the elements in the Lunar Return to see how this really "big picture" planetary aspect works for you.

Uranus Square Neptune

This several-year period (most recently centering around 1953) occurs only twice every 172 years and is reputed to have the effect of spiritual and scientific conflict upon society at large. Too lengthy to have much differing effect monthly, it nevertheless sets the tone of the period. Uranus/Neptune's relationship to the rest of the chart by house and aspect give it a reinforcing effect that is both inspiring and soothing, wakening and tranquilizing. So, look to its relationships this month to the rest of the elements in the Lunar Return to see how this really "big picture" planetary aspect works for you.

Uranus Conjunct Pluto

This several-year period (most recently centering around 1966) occurs only once every 128 years and is reputed to have the effect of transforming and sometimes violent technological progress upon society at large. Too lengthy to have much differing effect monthly, it nevertheless sets the tone of the period. Uranus/Pluto's relationship to the rest of the chart by house and aspect give it a united, two-for-one punch that is a regular monthly shakeup. So, look to its relationships this month to the rest of the elements in the Lunar Return to see how this really "big picture" planetary aspect works for you.

Uranus Opposite Pluto

This several-year period (most recently centering around 1907) occurs only once every 128 years and is reputed to have the effect of transforming and sometimes violent technological and political progress upon society at large. Too lengthy to have much differing effect monthly, it nevertheless sets the tone of the period. Uranus/Pluto's relationship to the rest of the chart by house and aspect give it a united, two-for-one punch that is a regular monthly shakeup. So, look to its relationships this month to the rest of the elements in the Lunar Return to see how this really "big picture" planetary aspect works for you.

Uranus Sextile Pluto

This several-year period (most recently centering around 1996) occurs only twice every 128 years and is reputed to have the effect of uniting political power and technological progress in society at large. Too lengthy to have much differing effect monthly, it nevertheless sets the tone of the period. Uranus/Pluto's relationship to the rest of the chart by house and aspect give it a supportive underpinning that calms the regular monthly rhythms. So, look to its relationships this month to the rest of the elements in the Lunar Return to see how this really "big picture" planetary aspect works for you.

Uranus Trine Pluto

This several-year period (most recently centering around 1922) occurs only twice every 128 years and is reputed to have the effect of uniting political power and technological progress in society at large. Too lengthy to have much differing effect monthly, it nevertheless sets the tone of the period. Uranus/Pluto's relationship to the rest of the chart by house and aspect give it a supportive underpinning that calms the regular monthly rhythms. So, look to its relationships this month to the rest of the elements in the Lunar Return to see how this really "big picture" planetary aspect works for you.

Uranus Square Pluto

This several-year period (most recently centering around 1933) occurs only once every 128 years and is reputed to have the effect of transforming and sometimes violent technological progress upon society at large. Too lengthy to have much differing effect monthly, it nevertheless sets the tone of the period. Uranus/Pluto's relationship to the rest of the chart by house and aspect give it a revolving punch that is a regular monthly shakeup. So, look to its relationships this month to the rest of the elements in the Lunar Return to see how this really "big picture" planetary aspect works for you.

Uranus Conjunct Ascendant

Uranus rising gives this month a brusque, edgy quality that may lead you to seem overbearing or too quick to judge when all you really were expressing was excitement and self-confidence. Original ideas may indeed abound, but couch them in terms that people can more easily accept and refrain from overwhelming them, which could lead to confrontation and failure to communicate. Don't be short or impatient, but bristle with enthusiasm and take the time to explain more than once if need be. It takes people time to absorb the new, and there's no point in scaring them off by coming on too strong. Also, objects may come out of the blue, so watch out when you're crossing the street, etc.

Uranus Opposite Ascendant

Look to others, especially a close partner, for inspiration this month, as they may be a font of new and original approaches to life and circumstances. You may find them somewhat abrupt in explaining themselves, but that's impatience born of the fact that they understand what they're telling you but you don't. Avoid confrontation while getting the details of what they have to say, and don't take it personally; it's only intended to illuminate and help. You should not only forgive personal eccentricities, but examine them and see if there isn't something in it for you, too. What may seem to be a strange or unorthodox approach may be just the thing to yield pleasure and profit, once you really get into it.

Uranus Sextile Ascendant

You will find that you can approach others and sell them on the most outlandish ideas—hopefully good ones—if you disguise them in the cloak of normality. The sly approach to slipping in the new dodge as if it were just a variation on a new move is just the ticket. Make people think they thought of it themselves, and they'll spring for nearly anything, but don't press it too far, as you should share in the action, too. Quirks of friendship and sudden income surprises both are in the cards, but you will have the ability to smooth it all out and take the shock out of the unexpected. By resonating with the situation, you use yourself to redistribute the energy, allowing all access with ease.

Uranus Trine Ascendant

You will find that your most inventive moments this month come when you just lay back and let the Muse in. Don't work on it; play with it. Don't try to nail down the details; pick up on the essence of the whole shebang. No matter how odd it might seem at other times, your most independent and original approach will sell without you having to sell it. When you are acting as a conduit for the light, the light shines brightest and clearest. Keep an independent eye on the process, however, so you can remember what you did right and can develop it more later. You have the ability to ease the way for others who would otherwise have a rocky time of it. Break the rocks into gravel, and you have a solid road.

Uranus Square Ascendant

Sudden events can make it difficult to chart a steady course this month, and if a sudden, unexpected gust hits you, expect another one. Choppy waters abound, but you stabilize the bouncing boat by staying in the middle seat, avoiding sudden moves that may capsize you. You'd like to get up to speed, but stop-and-go traffic has you testy. What first seems contrary, though, will even out in the end if you don't overreact, even though others may. Unexpected events can tar your reputation, shake up your house, but take them in stride, and, most of all, don't take it personally.

Uranus Conjunct Midheaven

Uranus overhead can mean that some strange things may be said about you this month, and it may be necessary for you to take steps to interpret them to your benefit. Both the most original and the most unorthodox parts of you will get air time, so be ready to spin the former as genius and the latter as cheerful eccentricity. Similarly, sudden changes and unexpected turnarounds may need to be fielded in the career department, so be ready to reverse your field and take off in a new direction when the opportunity arises. As the air crackles with your news, remember that flexibility is essential and trying to stand pat will be a losing gamble.

Uranus Opposite Midheaven

Unexpected events at home can catch you off base and distract you from professional matters, so be ready to tend to your personal flock when they call. Avoid confrontation over ethical dilemmas, as new situations demand new solutions and one size does not fit all. A new internal approach may be necessary in order to let differing points of view live in harmony. On the other hand, strokes of brilliance can bring solutions to living problems, and rearranging your surroundings on a whim may be just the thing. Doublecheck fuses and other electrical items that may fail or cause a hazard. If a confined space makes you crazy, don't squawk; take a walk and let the fresh air take care of it. The change of scenery will do you good.

Uranus Sextile Midheaven

Taking the unexpected and the humdrum with equal aplomb can gain you the reputation of being a problem-solver who comes up with original solutions to otherwise knotty tangles. By drawing on hidden inner resources that bubble up on cue, you can bolster who you are and where you are going. It might be a good idea to keep your resources under your hat, however. If they know the rabbit's already in there, the trick doesn't work very well. When you draw from a hidden well, it seems to others that you tap their springs—which, actually, you do, as we all share the inner life. When you have a novel solution, sell it as a joint illumination, and you'll get a brighter light.

Uranus Trine Midheaven

This is a good time to remind everyone of all those brilliant ideas you had that they (and maybe you) forgot about, as this month the air will carry the word afar. Stick with what you know that others don't, and you will be a seeming font of wisdom and originality you didn't know you were. Mine your inner resources for what is already there and advertise—you'll be a new product, and in demand. Don't make the mistake of putting yourself on automatic pilot, however, as now is the time when you'll want to be mapping the landscape you cover in a new way for use later. Stay alert, and when you hear you're right, you probably are, so bank on it.

Uranus Square Midheaven

Getting into tiffs with a partner or being too sure of your latest flash can land you in trouble that you'll have to untangle down the line, so tread with a little more reticence and care than usual. Sudden, unexpected moves do not increase the confidence of others, and you can become thought of as erratic, even if you are quite sure of yourself. Thus, don't shoot from the hip, either to initiate or return fire, and then if you don't hit the mark, don't waste the rest of your ammunition on a moving target. Harsh remarks can seem hurtful, but later you may find them well intended and, in the end, helpful. Don't take punches personally; there are many rounds to go.

NEPTUNE ASPECTS

Neptune Conjunct Pluto

This aspect has not occurred in recent historical times and will not repeat in a very long time, so commentary would be entirely speculation of no real value.

Neptune Opposite Pluto

This aspect has not occurred in recent historical times and will not repeat in a very long time, so commentary would be entirely speculation of no real value.

Neptune Sextile Pluto

This rare but lengthy outer-planet dance happened throughout the 1950s and again in the 1970s and puts a deep background on a generation when spiritual issues are not forced but sort of proceed in a leisurely matter without conflict, either because of mutual respect or blissful unawareness. So, actually, although this aspect has nothing to do with an individual Lunar Return, it sets a tone of tolerance, and this mellow relationship between Neptune and Pluto will manifest itself as a go-between that eases other aspects they take from other planets and angles that are active on a monthly basis. If the faster planets giving aspects are in trouble, this sextile eases the pain.

Neptune Trine Pluto

This aspect has not occurred in recent historical times and will not repeat in a very long time, so commentary would be entirely speculation of no real value.

Neptune Square Pluto

This aspect has not occurred in recent historical times and will not repeat in a very long time, so commentary would be entirely speculation of no real value.

Neptune Conjunct Ascendant

Neptune rising may make this month a somewhat foggy time when it's hard to pin things down and you probably shouldn't, anyway. That is partially because you are hard to pin down, which gives you great camouflage if you can use it, but isn't ideal if you are trying to make yourself perfectly clear. It is a good time to work on getting your fantasy world out into the open and making some of it come true, a possibility made easier because you will more easily fit into other people's fantasies, thus making it mutual. You can fool people with greater ease, but also risk fooling yourself. So, this aspect is suited ideally for pursuit of the spirit and the generally ineffable, which will be closer to your grasp.

Neptune Opposite Ascendant

This is a great time to find your dream partner, but not necessarily an ideal period for exchanging vows. You are likely to see others through a soft filter, which can make them look better than they are, but the illusion may fail later on, so enjoy it while it lasts. Because of this, you may be more easily subject to deceit, so don't let anybody pull the wool over your eyes in areas you'd rather they didn't. Similarly, expect others to be a little on the confused side now and then, so be patient if you find yourselves at sea in a fog without a chart. Sometimes it's fun to be clueless, and the mist will lift soon enough, baring a starker reality.

Neptune Sextile Ascendant

Expect the financial outlook to be a little mushy this month, but that allows you to fudge things to please everybody, a path worth taking because it will serve you well. To a certain extent, you may find that by taking on the persona of other people's hopes and illusions, you can help lead them to a better understanding of themselves, bringing you along in the process. So, feel free to empathize, sympathize, commiserate, console, and generally share mutual feelings that lead to better relationships and understanding. By serving as a mentor and a platform that helps others achieve their aspirations, you do them and yourself a favor. Right now you will play that role more easily than usual, so take up the mantle if the situation suggests it.

Neptune Trine Ascendant

This can be a strongly creative period, during which you pull down inspiration from the Muse, but don't try to get down to specifics or put the final touches on what you begin, as that will waste some of the energy that's offered. Stick to generalities, speculation, blue skies, pipe dreams, and what-ifs, and you may find that what you dream up might actually get put into motion later. It's probably wise to avoid legal specifics, as the fine print may be hard to read. Children are up to their usual tricks, and don't be surprised if they pull the wool over your eyes in the nicest possible manner. You may be the center of attention, but not even notice it.

Neptune Square Ascendant

It may be harder than usual to make yourself perfectly clear this month, as what you see is not what you get, and what you say may be totally misread. Attempts to clarify are likely to make things worse, so when in doubt, try another tack entirely rather than reexplaining until the cows come home. This may be because you don't look like what someone expected, or because you're not clear as to how to really express what your bottom line is. Chemical solutions don't help—in fact, they just confuse things more, unless you just decide to put it all off and party. Still, there's the hangover to consider, literally or figuratively. Time will put you back into focus soon enough.

Neptune Conjunct Midheaven

Neptune overhead flies like a chameleon flag and lends you the ability to disseminate whatever stories you like about yourself and likely be believed. The hardest one to sell, strangely, will be the actual truth, so pick a flag you're momentarily happy with. This same effect can muddy career matters where clarity is important and you'd like to be able to cut through the hype. Nevertheless, there are uplifting effects wherein yarns are spun about you that may be unlikely but can make you larger than life, especially if you're in a business that utilizes a lot of advertising, entertainment, or spiritual pursuits. If you want to maintain the illusion, however, avoid personal appearances, which might puncture the balloon.

Neptune Opposite Midheaven

You may feel a certain softness of purpose that makes you turn inward to refocus on what your inner, personal bottom line is. That can mean taking a little time for meditation at home, reestablishing goals and dreams you cherish that may have become neglected over time and need a little fresh paint. Although this may have the short-term effect of slowing down professional activities, in the long run it will shore up your confidence and resolve, fueled by freshly shaped aspirations. Take some time to relax and let the mind wander, without specific pressures to produce or change, and changes will evolve and congeal of themselves by month's end.

Neptune Sextile Midheaven

Indefinable inspirations that arise from the hidden places of the heart and mind gently reinforce your ability to make your mark on the world. Learning to lead by simply accompanying another is an important skill at this time and will do much to bolster your reputation. This is because when others see in you a reflection of themselves, they are most likely to give you the kind of support they would give themselves. Share your dreams, share their dreams, and life becomes a dream. Help, and you shall be helped along the way not by forcible promotion but by the quiet word that you are on the right side, pursuing the right goals, part of the team.

Neptune Trine Midheaven

Your strongest support now may come from the groundwork you have laid among fellow workers, teammates, and those who have put in their two cents with yours along the road of life. This may be mainly moral support and a thoroughly good word, not a cash reward, but it will represent a certain level of admiration and spiritual brotherhood that will only help you get ahead. You don't need to promote yourself, just let it happen when it happens, but be sure to get a record of it when it does. Recommendation and praise from your peers is one of the surest ways to maintain your reputation and move ahead in your professional or personal endeavors.

Neptune Square Midheaven

Keep an eye out for misreadings of who you are and what you've done that may be drifting in the wind. These may not be intentionally misleading—only wrong information passed on and magnified like a game of post office. Still, corrections will be in order, as conflicting stories about your intentions and your accomplishments will only muddy your professional efforts and dilute your reputation. These may have arisen as a result of conflict between your personal presentation and what you put out on paper or on the airwaves, so you may have to get your public-relations operations straightened out to match. Don't push it right now, just note what has to be done.

PLUTO ASPECTS

Pluto Conjunct Ascendant

Pluto rising this month can put an emphasis on personal confrontation and a general inclination to use force of personality to gain ends instead of negotiation or compromise. It can also mean that circumstances hold you down in a way you cannot get around and you must wait for them to end or go someplace else where it's not happening. The theme of both is the same: power used as a suppressant in order to get your way, whether you're doing it or it's being done to you. This is something to use sparingly on your end, in order to avoid reprisals from the disgruntled later on. When it happens to you, don't waste too much effort in resistance; move on to another game.

Pluto Opposite Ascendant

You may experience a squeeze play from your partners, either in business or pleasure, and the first tendency may be to refuse to budge and put up a fight about it. That, unfortunately, will just makes things more intense, and you may find it the better part of valor to yield to another's insistence, if only to be considerate and polite. The need to control comes out of insecurity, and to the extent it raises your hackles, you're equally afflicted. The best way to deal with a bully in the playground is to ignore him or, that failing, call in an air strike from the teacher. So it is later in life, where avoidance is the first best option, and alliance with the even more powerful the second.

Pluto Sextile Ascendant

You will not need to be overbearing in order to force a situation if you rely on your personal resources to do the work for you. Trusting in close allies will only make them closer, and the well refills when you draw from it. Subtle imposition is the order of the day, done in mutual interest and not setting aside one side for another. When resources appear to fail, you are the one to wake them up again, restart the engine, and encourage everyone to get back into the game and get on with it. This is best done by a personal appearance, where your force of personality can spark action, but where you are not burdened with being the only one being the bad guy.

Pluto Trine Ascendant

You may find that you have tremendous resources of well-placed faith when you have to overcome serious obstacles, and simply knowing you can do it will make roadblocks crumble and opposition flee. The faith may be in the deity, your friends, your background, or any other resource that you know is firm and will come to the rescue. When you demonstrate this, you will find that others will gain faith in you and look to you as a bulwark to see them through. When you decide, be utterly without doubt, full of the knowledge that you are right and a winner. When you do, you will find that the necessary help flocks to you like a magnet and you will pass with ease.

Pluto Square Ascendant

Fighting will get you nowhere this month, especially because that may be the first thing you're inclined to do, especially when roadblocks appear at work or home and you feel that bulldozing them will take care of the problem. It won't, and the more you try to force your way, the more resilient and elusive the resistance becomes, until you reach maximum frustration and minimum forward motion. The reason is that the people involved are at cross-purposes and not really hearing what's being said, but pushing on, anyway. The solution is to cut the spiraling shovefest and find out what the real issues are, after which you'll likely find everyone was really in accord to begin with.

Pluto Conjunct Midheaven

Pluto overhead this month can either help or hurt your career prospects, depending on how you use it. If you allow blockage on the professional level to hold you back, you'll have to bide your time until it passes. If you let everyone know beforehand that they're dead if they cross you, you'll have it made. You don't actually have to be ready or inclined to be forceful, but others should hear that you are and, as a result, back off. The result can be a regeneration of your reputation kicked up a notch. Avoid being tagged as a roadblock, rather as one who is powerful in your field and willing to exercise that power. The best defense is what others see as a really tough offense.

Pluto Opposite Midheaven

Expect a little resistance on the home front, where there may be overlooked stones in your pathway that get in the way and refuse to move. This can come from a person, an article in your environment, or just a general feeling of not being entirely safe. To use it to your best advantage, don't debate the specifics but plunge to the foundation of the matter—the motives, the feelings, and the hidden agendas that may be fundamental to the situation. This is a judgement call, however, as you don't want to be dismantling or replacing something intransigent if it's going to be leaving anyway. The only time to force things is when your bottom line is brought into question.

Pluto Sextile Midheaven

While you may have to be up front about what you do, you don't have to give away the sources of power that help you do it. What or who is hidden or buried in the background can be your best support and ally this month, especially if you are discreet and don't give them away. It is important to be seen as the one on top through your own efforts alone, while employing silent partners to give you an extra push when you need it. Pulling strings can get you more than confrontation, and it may be worth being in debt as a result if it achieves your goals. Underground supply lines may be just what you need to put you over the top and make it seem like a breeze.

Pluto Trine Midheaven

You should be relatively free from the slings and arrows this month, though should your reputation be challenged, you will toss it off your back with ease. The essence of this ride is being supremely confident that the work you have done, the accomplishments you have made, and the people associated with you in your efforts are unassailable and will come through any gale unscathed. The very fact that you feel this way and show it will in itself ward off provocation, though you should by no means use this as an opportunity to slough off. Rather, use your position as a platform on which to build more of the same, utilizing the support already in place for leverage.

Pluto Square Midheaven

You may be subject to some assaults on your reputation, possibly born of jealousy or envy, that may be hard to ward off, as they can be of a repetitive nature. It's probably not a good idea to try to thwart them head-on, as that will only exacerbate things—better to find out why those involved are at odds and how to correct the situation. Similarly, you may experience some serious jostling in the career arena, and both time and energy risk being wasted. If you know what you've got is beyond reproach, stay out of the fray. If not, try to draw attention away from yourself and thus deflect any possible real harm.

NORTH NODE ASPECTS

North Node Conjunct Sun

The chance to undertake new responsibilities is open to you this month, and your initial reaction will be to go ahead when the occasion is presented. This could be a business or career deal, or it could mean a new involvement with another person who will in some way depend on you. It seems like a glowing opportunity and a creative outlet for your energies, but remember that you will likely have to continue with it when the glow is gone or when you have less energy to burn, so look at it in perspective. If it will be a recurrent joy and a lasting reward, you have everything to gain for your commitment. If any aspect of it is truly temporary, look again.

North Node Opposite Sun

You may find yourself in receipt of gifts and affections that are quite unexpected and that may seem to be more than you deserve. Think that if you will, but don't say it. Free gifts are payment for good deeds done that you may not even remember or think connected. Remember "What goes around, comes around"? Well, here it comes, so enjoy your reward, even if you don't know what it is for. We're not necessarily talking cash on the barrelhead here, but perhaps special moments of love and beauty that don't happen often, relief from previous burdens, or just the glowing realization that there is more joy around you than you might have thought. If it brings happiness, it's a gift.

North Node Conjunct Moon

Emotional attachments this month may last longer than you think, so when you're asked for a favor or commitment, it could be more serious than you might imagine. Remember, it is easier to say you'll do something than to actually do it, especially if you have to do it repeatedly, and although the initial feeling may be great, the consequences may not be. Nonetheless, that first take on the situation may be the most insightful, and seizing an opportunity may depend on reacting quickly, leaving you with a choice: think it over or go with your instincts. Knowing ahead of time that this may be coming will allow you to be more prepared, so tune up your antennae and polish up your judgment faculties and you'll be ready to make split-second decisions.

North Node Opposite Moon

An emotional gift likely awaits you this month, a moment when you feel you are getting some kind of payback, some response and acknowledgment from the universe to who you are and what you have done. This could be a case of recognition from your peers, an instance of special affection from someone close, or even just a strong sense of déjà vu—that you're getting some kind of reward for something you might once have done, somewhere. The moment could be fleeting, but if you catch it, it can be a memory you'll keep with you for a long time. Though this may sound crass, make sure you've got a camera or some kind of recording device to capture the moment.

North Node Conjunct Mercury

A really bright idea this month could put you in clover—or in hot water—depending on your judgment call. Something you dream up or a clever scheme someone else lays on you can offer you the opening to get in on the ground floor if you are willing to make a commitment. This may be just the chance you were waiting for—or it could be the opportunity to wear a millstone around your neck, once the novelty wears off. Therefore, without being impolite, do look a gift horse in the mouth, and look before you leap. Consider the impact your ideas will have if you act on them—and later are forced to 'fess up to them. Make sure you will get to take credit in the end, not pay off the debts.

North Node Opposite Mercury

Messages from the past, from old friends or from earlier scenarios, make take you somewhat by surprise. Or, you may receive the simple gift of a great idea from someone who asks no credit for it. However it works, because you aren't sure of the pedigree, you may give it the jaundiced eye. But there's probably little need to, as you are definitely on tap for a bonus prize if you keep your eye out for it and don't let it flash by unnoticed. Make sure there really are no strings attached, and then go ahead and use and enjoy it with a little thanks to the folks upstairs. The fact is, you've done something to deserve it someplace else, so treat it like you've earned it.

North Node Conjunct Venus

The opportunity to get your hands on something or someone you really want, but at a price, is on the horizon. "At a price" is the operant phrase, and the ultimate price may not be immediately evident. What looks like a fair deal for the moment may have you on the ropes down the line thanks to the fine print, so don't just scoop up anything you like without a second glance. If it really looks good on second viewing, then go ahead, as you won't likely get a second chance. They say he who hesitates is lost, but here read it as he who looks hesitant loses. Keep your wits about you and don't appear indecisive, but don't get taken for a ride. Those caveats aside, go for it.

North Node Opposite Venus

Look for some manna from heaven this month, and if you don't immediately see it, look again. It could be in the form of a wonderful new relationship, an unexpected injection of cash, the chance to acquire something you really wanted, or a gift of some kind that you didn't anticipate. Whatever it is, don't look this gift horse in the mouth, just accept it thankfully and devour it. It might be in the form of a favor returned, and in any case, that's just what it is. Every now and then it's your turn to take, because you were giving another time. It's not always a return transaction between the same two people or situations—it's just a way to let you know the eye is on this sparrow.

North Node Conjunct Mars

Watch carefully that the actions you take this month don't have unanticipated or unintended effects that can catch up with you later. Realize that taking action on any matter implies your commitment to it, at least in part, so be sure that what or whom you associate with has a clean bill of health and will not come back to bite you. That doesn't just mean don't be rash—it means be extra smart even about mundane or routine matters, as they now can develop unnoticed hooks that can drag you in directions you might not mean to go in. On the other hand, you'll also find that when opportunity knocks right now, it will require concrete action. Just make sure the concrete is mixed right.

North Node Opposite Mars

This is a really good time to make sure you've got your back door covered and you didn't leave a window open for the rain to get in. Inadvertent moves, or actions taken by others on your behalf without your knowledge, can cause setbacks, so keep your guard up a bit more than you ordinarily would. The energy now can also put force behind you that boosts you crazily along, but without sufficient control to assure a safe ride. If, despite all precautions, you take a hit from left field, just tell yourself that it's one more karmic debt you don't have to pay off. Sometimes it is necessary to be in receipt of the mistakes of others in order to see our own.

North Node Conjunct Jupiter

This next month or so can feed the inclination to go whole hog and seize the day by jumping in with both feet. That can apply to personal as well as career enterprises, which can be a thrill and a challenge. Remember, however, that big ideas and promises entail big delivery down the line, so you had best scope out the road ahead before climbing on a bandwagon. That said, you'll also likely see more than your share of optional new directions asking for your commitment, and if you're ready, there's no time like the present. Don't let life pass you by. The new shoe you slide into today may be tight on you tomorrow, however, so leave yourself some room to grow.

North Node Opposite Jupiter

If doors suddenly open and your universe expands for no particular reason, hop on and take a ride. This could be a time of gratuitous, unasked-for learning experiences that are fun to live through and great to have in your back pocket later. This means that if something new beckons, with no strings attached, it's like a free course in school. All it takes is your time, and it is its own reward. New company, new surroundings, new pursuits to enjoy—open your horizons wherever possible as you are very likely to have the chance to change now, and your ability to absorb and incorporate the new into your life is expanded.

North Node Conjunct Saturn

For several months the opportunity to take on burdensome projects and responsibilities is in the air, and it is up to you which, if any, to choose to take on. Choose wisely, as almost anything you sign up for now will last longer and require more effort than you imagined at the beginning. Similarly, events occurring around now have a certain gravity that makes you take them more seriously, and usually for good reason. This can be for long-term commitments, but it also can mean that short-term undertakings can turn into long-term woes and a brief fling could dog you for a long time to come. Don't be paralytic about it, just watch what you're getting into.

North Node Opposite Saturn

For a couple of months it might be wise to be wary of unintended or unwanted snares that can bog you down. That means test the ice before you step on it, put your toe in the water before you dive, and listen to the weather forecast before you take a trip. You may do well to chalk up delays and side trips to something you did in another life, and get on with things. Still, you can be aware that juicy morsels of bait may have hooks in them and silver linings have clouds inside them, and thus absent yourself from involvement wherever it is suspect. Common sense will spare you when cleverness cannot. You can't escape a little drudgery, but you don't have to deliberately walk under ladders.

North Node Conjunct Uranus

You may be faced with a rather sudden decision of whether or not to participate in a new involvement, something or someone who comes out of left field. It may surprise or even shock you, but it can also have a breakaway appeal that just might be the thing to dive into. Chances are you won't have much time to think about it, but even in the haste of the moment be aware that it will involve more than you think and may put you either on the leading edge or simply the outer edge. If you are willing to take up something you can, and will need to, justify later, you may have a chance to achieve personal and professional breakthroughs. Just be ready for a bumpy ride.

North Node Opposite Uranus

Rocks, shocks, and potshots from left field are in the atmosphere for a bit, so keep your eyes open for something unusual coming out of thin air. This may come in the form of unusual or surprising behavior from someone you think you know, or it may be rapid developments in an unexpected area that bowl you over. Whatever it is, the ball will be in your court and you will have to return it. It might not be a bad idea to reexamine whatever seems especially normal around you, as by nature that's where surprises come from. The more open-minded and centered you are, the more likely it is that you will benefit rather than be upset by sudden developments; utilize their energy rather than being derailed by them.

North Node Conjunct Neptune

For a month or so you may find yourself being asked to commit to things that really aren't clear enough to tie yourself down to, like promises to always be true, sort of, or a hot new project, but no contract. Unless you can make a commitment that is as ambiguous as what is being offered, it might be a better idea to say thanks, I'll take a rain check. This doesn't mean opportunities right now won't pan out, just that you don't know how they'll develop later when you're still tied to them. On the other hand, if you can make it a definite maybe, you'll be in gravy and not in the soup no matter how you play it, and you'll have had a good time to boot.

North Node Opposite Neptune

You are in a general climate of uncertain rewards right now, and while you think you know what you are getting into, you may not. Alluring promises can turn out to be hot air, while assumptions you make give way beneath you. There is an element of shifting sands that is unavoidable, but you can be on steadier ground if you question much of what you hear (don't disbelieve, just question), and don't be disappointed if it's not really what you thought. Take the usual precautions against catching a bug, and make sure you have a designated driver, just in case. Favor activities that don't require a high focus and that you can relax into.

North Node Conjunct Pluto

This is a time when forceful commitments are made, sometimes out of great conviction, sometimes out of a lust for power, and sometimes out of nothing but fear. That's the same for everybody, and the focus of matters can be a little more extreme than they warrant. A general rule of thumb is this: If you feel you must have it, you probably shouldn't; if you feel you're absolutely correct, you're not; if the situation seems totally scary and overwhelming, back off and it won't be. Extremism on behalf of almost anything is a mistake, and it's one you'll have to live with a long time if you make it. So fight the tenor of the times and refuse to sign the pledge, whatever the pledge might be.

North Node Opposite Pluto

Squeeze plays can happen where and when you least expect them right about now, and you are equally as likely to be the perpetrator as the victim. It's in the air. It's a time when you may get what you deserve—for good or ill—and not be able to do very much about it. Well, if you deserve it, why should you be able to do anything about it? It's a time to learn to bow out gracefully and not play the game anymore. Passive resistance works better than fighting the tide, and looking for new options to permanently change things is ultimately the way to go. If you have to make a choice that is not truly yours, document it so you can prove it later. Turn a shove aside into a push forward.

North Node Conjunct Ascendant

As a general rule, in your personal life you will be more on the giving side and your partner(s) more on the receiving side this month. This means taking a load off their shoulders, volunteering to be the first to do the job, and serving as the forward column that takes the brunt of it to protect the rear. You will do this voluntarily if it is to be of any benefit to you, but you'll probably get wangled into it like it or not, whatever your wishes. There's no point in putting up resistance, just adopt the role and play it with a will, as the shoe will be on the other foot soon enough. And, if you learn to like it, or like it already, it will be hard to give up.

North Node Opposite Ascendant

As a general rule, in your personal life you will be more on the receiving side and your partner(s) more on the giving side this month. If you're not entirely comfortable with that role, now's the time to learn it well. If it's already your fave, relax and enjoy it. Don't push for more than you think you deserve, and you will get more than you think. Be demanding, and you will find that when the shoe is on the other foot, as it will be soon enough, it will be a tight fit, indeed. Professionally, in a competitive space, this month may leave you somewhat open to gratuitous hits, so make sure that when you are collecting your dues, they are dues, not comeuppances.

North Node Conjunct Midheaven

This may be a more important month for your career efforts than you know, as responsibilities and new endeavors taken on now will be tied to your reputation for a long time to come. That means that special opportunities present themselves, but so do occasions for your undoing if you are not wise enough to differentiate the two. So, whatever your choices, this is a good time to be thinking long-term and not business as usual, as you'll be filling out your resume and maybe lots more with this month's engagements. The one thing to avoid doing is nothing, as circumstances with this potential aren't an everyday occurrence and are a gift not to be refused.

North Node Opposite Midheaven

This month could be a period during which you will have to depend on the vagaries of fate and the goodwill of your associates to defend and secure your professional position. Your flag is up the pole, and you can't control who salutes it. The groundwork of friendship and support you have hopefully laid will now be called upon to vote on you. It's either pay time or payback time, or perhaps a little of both. Your more important field of play, where you have more influence now, is at home, where what you take up now can lead to general improvement. Just watch that you don't get too tied to it in a way that would interfere with your later outside operations.

Chapter Six
FEATURED NATAL PLANETS

Any Natal Planet Conjunct Lunar Return Sun

This highlights your fundamental energies that make your personality shine throughout the month and beyond. It will mean that events this month will to some extent revolve around your ability to demonstrate self-confidence and become a well of energy and inspiration for others. It is the opportunity for your self, the unity behind the sum of your parts, to take command and establish power and influence over your surroundings. That does not mean that you will be lord of all you survey, but it's a sure step in that direction and will lead to confirmation of an expanded self-image that can only put you on a rising track, both on the inside and the outside.

Any Natal Planet Conjunct Lunar Return Mercury

This month will tap your mental energies in such a way as to make what goes down highly interconnected with and dependent on your ability to express yourself clearly and get your ideas across effectively. It will be both a challenge and a privilege to have your point of view valued and validated by greater exposure and recognition. At the same time, expect more challenges to come from enhanced exposure, so you will need to have your arguments lined up and ready to explain when questioned or tested by naysayers. Positions you take now may well have to sustain you as the center of longer-lasting efforts, so see that you build your foundations on rock, not sand.

Any Natal Planet Conjunct Lunar Return Venus

You will likely run across issues and opportunities that allow you to demonstrate your charm and magnetism and set the tone for some time to come. Expect to have both the need and the desire to captivate your audience and come away with the prize due to a combination of innate charisma and studied attraction. Getting what you want, be it money, love, position, or more altruistic gains, will be highlighted. So make sure that you not only get what you want, but want what you get, as you may have to be satisfied with it for some time to come. Be careful of what you wish for, as you just might get it, and when you do, you will want it to satisfy for the long-term.

Any Natal Planet Conjunct Lunar Return Mars

Your physical energy and get-up-and-go are put in focus this month in a way that showcases your ability to sustain efforts and makes an impression of robust strength that engenders admiration and trust. It means, however, that you will be committing yourself to a rhythm that you will likely have to repeat, so don't design a regimen that is beyond your capabilities or will leave you drained if it becomes the expected routine. Hit hard and fast when needed, but don't use up all your ammunition on the first target you see. The essence of exercise is that it should not exhaust you, but should reinvigorate and leave you stronger for the effort. So it is with life.

Any Natal Planet Conjunct Lunar Return Jupiter

You can expect your innate inventiveness and ability to expand upon new opportunities to get special focus and a chance to shine. What happens this month may well stick with you for some time to come, so be sure your schemes are solid, not harebrained, and your inventions practical, not frivolous. Originality for its own sake would be a flash in the pan, but innovation integrated with sensible implementation will leave a legacy that has your name on it. You don't need to nail things down chapter and verse, but the overall plan should be solid and unshakable.

Any Natal Planet Conjunct Lunar Return Saturn

Your ability to do less with more and come out with lasting results will be put in the spotlight, as will be your capacity to sustain your efforts and exercise authority within your personal territory. If you appear to have your act together, this month will be remembered as the time you demonstrated it and will enhance your credibility for some time to come. Thus, in a unique way, challenges become opportunities, restrictions become bulwarks, and working with less means being able to travel lighter and faster. Don't be afraid to be seen as a taskmaster, either to yourself or others, as the extra discipline you show now can make the month, and months beyond.

Any Natal Planet Conjunct Lunar Return Uranus

Broken-field running and unusual and spontaneous reactions to unexpected events can become your calling card and something to be remembered for. This offers the opportunity of praise and admiration for thinking on your feet and responding with original solutions to sudden situations. It also offers plenty of chances to stumble and trip, so tread lightly and watch your footsteps. Be nimble, be quick, and make your operations stick. Avoid pushing, insistence, intolerance, and impatience, as those are the greatest slipping points. Choose your shots, and don't shoot until you see the whites of their eyes. One strong, well-placed effort will be all you need.

Any Natal Planet Conjunct Lunar Return Neptune

Your abilities to inspire and beguile will be put on display this month, so make a good show of them so it will have been worth your while. By leading with your dreams and ideals, you can induce others to follow suit, even though their goals may not be exactly the same. By putting an extreme spin on an otherwise untenable situation, you can turn people's heads and get them to accept what is best for the long run, even when it means short-term sacrifice. When you need to camouflage your efforts or mask your moves, you can naturally fade into the background until you are ready to surface at a time of your own choosing without fear of ill consequences. But know that how you do all this can easily become habit, so make sure you can live with it.

Any Natal Planet Conjunct Lunar Return Pluto

Your ability to seize control of the moment or wrest it from someone else can become a focus for the month, but should it be so, it will be remembered so. So, watch your step and don't step on anyone you don't mean to. The greatest force is the threat of force unused, and the most successful army never has to fire a shot in anger. Similarly, the greatest resistance is disappearance, so don't tangle head to head in a win-lose confrontation, just melt away to play next day. How you handle intransigent challenges and blockages right now can establish precedents for the future and habits that you will follow and others will expect. Don't deform, transform. Reform, don't conform.

Any Natal Planet Conjunct Lunar Return North Node

This month can highlight your ability and inclination to choose well and faithfully respect new commitments and responsibilities. It will probably give you some great choices to embrace, and some real clunkers to avoid. Remember that anything you do is likely to get you more embroiled than you think because it will make more waves than you expect. So don't rush into things, personal or financial, but when you do decide, do it body and soul. These opportunities will not likely go unnoticed, nor will your decisions concerning them, but don't let internal or external pressure cause you to serve up more on your plate than you can handle.

Any Natal Planet Conjunct Lunar Return South Node

This month can be a grab bag of unforetold events that can bring gain or throw up obstacles, all through causes and purposes that you are unaware of or have no control over. It's kind of like forgetting and leaving your back door open. The first person to breeze in may be the prize patrol or the tax collector . . . who can tell? Or, maybe, just some fresh air and a stray bug or two. In any case, you're more open than you think you are, so have backups at the ready in case of a systems failure, and have a pot ready for the gold should you suddenly find that the rainbow ends at your house. And, in either case, savor what it's like to live a little more out in the open.

Any Natal Planet Conjunct Lunar Return Ascendant

How you look and your physical presence will be highlighted at critical events this month, which means if you want it to happen, show up and don't send in a substitute or a note from your teacher. Get in there and get recognized, remembering as you do that you may be establishing a new groove that you can either dance to or get stuck in down the line. So spruce up, practice your moves, and get onto the floor where you can be in the thick of things. Steps you execute now you will be remembered for, at least until the next set, so step lively and step well. Put your best foot forward, and jump into the ring.

Any Natal Planet Conjunct Lunar Return Midheaven

Much may swing on what people say about you and the reputation you have gained, so be prepared to defend and promote your career position and your strength of character.

Moves made by yourself or others at this time may have especially long-lasting effects, as first impressions are often the most lasting, and most impressions are secondhand. This may remind you that your public image can be more important than your private one, and it is certainly more accessible to the majority of people. What they remember may not be the "real" you, but you will be wise to make sure the unreal version looks as good or better.

Chapter Seven

TRANSITS INTO NATAL HOUSES

Sun into Natal First House

This is likely your most interesting and active time of year, when you become more centered in the spotlight than at any other time. Take this month, therefore, to make hay while the sun shines, as it surely shines on you now. You can push yourself farther with more self-confidence and make more headway than twice as much effort would have brought six months ago. It's more than an ego boost, it's a demonstration of just how well you can do given a fair wind to work with.

Moon into Natal First House

This is likely your most interesting and active time of the month, when you can project yourself more convincingly and with greater force than at any other time. It's the time to get things done in person, face to face, rather than by mail or phone, as your ability to project yourself is at a peak, as is the inclination of others to accept you at face value. It's a period to make extra space in your schedule, as you're likely to have more to do than you expect.

Moon into Natal Second House

Attention to cash flow and how to improve its balance in your favor catches your focus, and you'll find yourself basing both your attitude and decisions on how much it costs, what it's worth, and who wants to buy it. Your innate possessiveness is at its maximum, and it may be hard to part with belongings, even ones you need to get rid of. It's also a time for reevaluation of what you think is important to you and why you consider it so; a time to get your ducks in a row and make critical choices.

Moon into Natal Third House

Getting the word out, making connections, and tying people and things together are the focus, and you can expect more phone and mail activity than usual, so adjust your schedule accordingly. It's a good time for touching base, but not ideal for in-depth conversations that may be put off until you have more time to get into them. Technical matters, networking, and anything that involves media are on the front burner, and machines can take on a life of their own, albeit briefly.

Moon into Natal Fourth House

A general inclination to hold back and go inside for inspiration holds sway, which can lead you to tend to home affairs or simply to withdraw a safe distance to consider what's important and what your next move will be. Similarly, it's a good time to refine defense strategies and patch up those fences that separate you from the neighbors and the outside world, literally and figuratively. When you are sure of your boundaries, you can more safely go beyond them.

Moon into Natal Fifth House

This is a very open time of month, when spontaneity and warm feelings can flow more easily, and the creative and romantic urges compete to hold sway. This is a good time for connecting with children or for taking a little time off for fun and frolic. Risk-taking is easier, and you're likely better at it now, though don't go out on a limb unless you're sure. Don't push for concrete steps, but go with the flow and let things happen and the results will surprise and delight.

Moon into Natal Sixth House

This period has a bit of a "back to work" flavor, where picking up stitches and nailing down projects is the flavor of the day. If you have gotten off schedule, either at work or in your personal habits or health regimen, now is the time to make the correction and get back on track. By strengthening routines so they run by themselves, you free your hands for more important things down the line and prevent untimely trip-ups from overlooked details at the last minute.

Moon into Natal Seventh House

Personal relationships can be in focus now, and you may find that you can share your emotions and responsibilities with a partner who will take a load off your back. By giving credit where it's due and avoiding competition or jealousy, you will find you often receive more than you give, but then that is what usually results from an open hand and heart. Fairness is at a premium right now, so go out of your way to see that everyone is on an equal playing field.

Moon into Natal Eighth House

Credit can be looked at creatively and used to maximum effect, with particular attention to raising your limits without overtaxing yourself. Recycling efforts may pay off by repurposing something you would have otherwise disposed of. A glance back to the old ways of doing things can inspire and even pay off as you realize there's nothing really new under the sun. Taking out the trash, once you have decided what it really is, can feel like a fresh, new start.

Moon into Natal Ninth House

This can be an ideal time for those long conversations and ambitious exploration of new mental and emotional territories. Your reach feels greater than before and you are only limited for the moment to what you can imagine, leaving the details for later. You can stand back and examine the big picture and chart your way through it. Travel broadens the mind, whether it be a physical trip or an armchair journey of the mind. Believe in what you conceive, and it will come true.

Moon into Natal Tenth House

Professional considerations should be looked at and adjustments made to see that your name is recognized and respected among your peers. While your attention is on it, touch up that resume, touch up your image, and take some time to advertise yourself. Get mileage out of compliments and take credit where it is due. Follow up on leads and don't be shy about blowing your own horn, as long as you can truly live up to what you promise. It's time to wave the flag—yours.

Moon into Natal Eleventh House

This is a good time to look to close friends for support, either as backing for endeavors you need help with or just for general emotional support and refueling. It's also a good time to tap into other people's creativity when your own inspiration well runs a bit dry, and you'll find yourself refreshed as a result. In a similar vein, you may also want to tap into higher resources, both spiritually and financially, to pick up some of their overflow for yourself.

Moon into Natal Twelfth House

This is likely the quiet time of the month, a lull in traffic, the calm before the storm of renewed energy that will follow. Retreat and contemplation, with resultant renewal and regrouping, are favored pursuits. Use down time to reconsider your options and make sure you are really on the right course. Inner challenges you grapple with and resolve now will result in more focused and effective action when the time comes for it, which will be shortly.

Chapter Eight
TRANSITS INTO LUNAR RETURN HOUSES

Moon into Lunar Return First House

This marks a period when it will benefit you most to use your personality as the winning card in any game you are playing. It's the time to make that sale, impress that client, charm that lover, or play to the audience. Don't try to impress them with statistics or your track record, just wow 'em with the way you come on, with the conviction you have in what you present. This is the month's window for being on-stage, so take the opportunity to light up when it's presented.

Moon into Lunar Return Second House

This month's finances are best dealt with, or at least planned for, during this two-day stretch. Make decisions, allocate funds, plan a budget, write checks—anything that can go towards wrapping up your money issues for the month. Once these issues are out of the way, you can move on to networking on your next set of issues. This is the time to count your recent winnings, impact any losses, and see what your bottom line is going to fund for you in the near future.

Moon into Lunar Return Third House

It takes a village to get anything done, which means to get things moving you've got to talk to everybody in town. If you take this time to be the great communicator and unify your social network, you will find there are plenty of hands to go around to get it all done. Open new avenues of discussion that focus not on theory but on everyday nitty-gritty, things that grease the skids for all concerned and avoid wasted time by getting rid of repetition and misunderstandings.

Moon into Lunar Return Fourth House

Attention to laying the foundations of what you are going to do and be for the month is the order of the day. Build from the ground up and don't start on the first floor until the basement is done. You don't have to complete it all right now, but have the order determined and your materials ordered so you don't have to go back and draw up altered plans later. Where you've already put it all together (or mostly), tidy up the joint, and set your house in order.

Moon into Lunar Return Fifth House

Making the most of playtime can benefit you as much as the most earnest work, and setting time aside for recreation and generally creative pursuits will serve to refuel and reinspire the rest of the month. The works of your life are your children as much as any son or daughter, and when they are conceived and nurtured in love, the result can only bring reward. So, listen to the natural soul inside and take some time now to bring those inner voices into the sunshine.

Moon into Lunar Return Sixth House

The daily structure of your life and work is often taken for granted and so can fall into disrepair. This is a time to reinvent the ordinary and make it special, to see that the habits and methods that make up your daily operations actually serve the purposes for which they were intended. By being a do-it-yourself efficiency expert, you can make this month more productive and less repetitious. If it works, embrace it. If it doesn't, erase it.

Moon into Lunar Return Seventh House

There is no greater opportunity in life than another person, and your ability now to partner well can turn this period into a cornucopia of possibilities that would not be available operating solo. Don't expect to luck into a good relationship—the best ones are carefully built and meticulously tended. Creating and maintaining space for both to operate comfortably in is the key to making a partnership more than just two people making do. Tend to each other's houses and hearts.

Moon into Lunar Return Eighth House

Financial self-reliance is a goal, but sometimes it's better to gamble with other people's money than your own. A carefully wrought juggling act of balancing borrowed stakes with retooling and refitting your own personal resources can serve you as well as gold in a safe deposit box. Extra baggage that you eliminate right now will make that whole process work more efficiently, so what you can't reclaim for reuse, spin off as gifts or collateral.

Moon into Lunar Return Ninth House

Is your life cinéma vérité, or is there a plan? Now's the time to invent the latter or improve on what you've got. Dead reckoning can get you where you're going, but why not have a map? This is a good time for establishing the lay of the land, getting that aerial overview that confirms where you've been, where you are, and where you're going. You needn't rush off on the journey, just turn it around in your mind until you have a good grip on the possibilities it offers.

Moon into Lunar Return Tenth House

What people believe you are often seems more important than what you really are, so crafting your public image is critical for success. This is always a mix of real achievements mixed with what's in the eye of the beholder, so make sure that you have both working for you. Find out now what other people, especially your peers, are saying about you and provide information to turn the opinion polls your way. Sometimes a spin doctor is more valuable than an M.D.

Moon into Lunar Return Eleventh House

What you can't do for yourself, maybe your friends can do for you, and a friend in need is a friend, indeed. You don't want to tap your nearest and dearest too often, but asking for help and support occasionally is part of what makes a good friendship, and you might do that now, even if you're not terribly in need, just to keep relations from getting rusty. A little mutual admiration can go a long way towards lifting your esprit de corps, and a heartfelt hug beats a handshake.

Moon into Lunar Return Twelfth House

Behind-the-scenes maneuvering can go a long way toward setting you up for the main event, so don't feel you have to tell all right now to get things done. What you don't know can hurt you, so some noninvasive information-gathering on the sly may be just what you need to spend some time doing right now. Honesty is the best policy, but do no harm is the bottom line. The less said the better right now, and actions will speak louder than words when the time comes.

Chapter Nine
Transits to Natal Planets

Sun Conjunct Natal Sun

This is the occasion of your birthday, so many happy returns! It is also the time of your Solar Return, the chart that tells you much about your coming year, just as this Lunar Return reveals highlights of the month. If you've got enough advance notice, you can even change some facets of the coming year by traveling to a different place for your birthday. It's definitely worth having a look at, and it will give you a larger context to better understand your Lunar Returns.

Moon Conjunct Natal Sun

This is likely your second most active time of the month, as your inner energies get a boost and you are more likely to be in for some ego-stroking. As greater energy and enthusiasm flow, so do the opportunities to expend them, so you might take care not to overdo things or you'll drain yourself. This has a "launch-pad" feel to it, so make sure new endeavors begun now are birthed with a steady hand, as well begun is half done and you don't want slips at the starting gate.

Moon Conjunct Natal Mercury

Sudden inclinations to figure things out and think things through are definitely the way to go, and it's a good time to put it down in writing lest you forget it. That's also a good idea because schemes hatched now may be somewhat colored by feelings of the moment that can be wisely edited out after a second look. Keep it simple and don't get bogged down in details for the time being, and you can cover multiple topics and issues and solve multiple problems while the inspiration lasts.

Moon Conjunct Natal Venus

Take the opportunity to express the warmth you feel and don't hesitate to display your social charms in a fertile group context. Flattery will get you everywhere, so you can safely lay it on thick. Self-improvement and beautification are the way to go, but self-indulgence can slip into the middle, so watch that you don't overdo things because you just can't help yourself. Impulse buying is easy to do—be kind to yourself, but watch your wallet.

Moon Conjunct Natal Mars

You may find yourself running into brief spells of irritability and/or impatience, when you want to get on with things but your way appears blocked. It's easy to get into a fight, and perhaps even easy to win one, but it's not a good time to choose one. Avoid quarrelsome and contentious people who might get your goat, and remember that most battles aren't worth fighting. Snap judgments are usually the worst, and you can't regret mistakes you don't make.

Moon Conjunct Natal Jupiter

Look for circumstances to further your good fortune and new ways to develop your agenda. This is aided by a generally positive, can-do attitude and a feeling that it's all going to work out for the best, and that's an approach that can make for excellent self-fulfilling prophecies. Generosity is the first option, which tends to be returned in kind, and a feeling of being in touch with the general operating system of life gives insight to the heart as well as the head.

Moon Conjunct Natal Saturn

If you are tempted to highlight your insufficiencies, do so in the interest of a better and less wasteful existence. When things suddenly stall, it's usually for a reason, and now's the time to find out why. Don't bash yourself, improve yourself. You can't overcome your limitations until you truly know what they are. Similarly, limitations are often imposed because you haven't made sufficient use of what's already on your plate. Say grace, and you'll receive grace.

Moon Conjunct Natal Uranus

The urge to think or act impetuously offers both risk and opportunity. A sudden change of direction or a new direction entirely can put you way ahead of the game, or get you thrown out of it. But, if it's within reason, fly with it and see what happens—nothing ventured, nothing gained. The best maneuvers often burst out full-blown like Athena from Zeus' brow. Don't be impatient with others who may not go along, and be ready to accept consequences, for good or ill.

Moon Conjunct Natal Neptune

Intuition runs strong for a bit, and chances are that hunch you have is correct, even if you can't put your finger on it or give a reason for your predictions. The same goes for your people-antennae, which are particularly tuned in to the feelings of those around you. It's easy to drift off into a daydream, or simply fog out at a cocktail party. Operating heavy machinery, however, is less than recommended. Don't try to dream and drive at the same time.

Moon Conjunct Natal Pluto

The inclination may be to push through to get your way, because for the moment the end seems to justify the means. Stand back and take another look, however, and you might decide otherwise. When you simply "must" do or have something, you probably shouldn't—at least not to the degree you are contemplating. Such feelings are signals that can get you in better touch with deeper emotions and are better used in that cause, rather than in the service of a quick ego boost.

Moon Conjunct Natal North Node

Commitments undertaken at this time may last longer and require more of you than you now anticipate, so don't sign on the dotted line until you have read the fine print. And, don't casually say yes to something you don't really mean, because you may be held emotionally or financially accountable for it. This is a time to finetune your honor and become more closely aware of your emotional responsibilities, and realize just how much your behavior affects others.

Moon Conjunct Natal South Node

Watch for small-time instant karma, bringing unexpected payoffs or well-deserved comeuppins. There may be a surprise package at your door, but read the note before opening and make sure it's not ticking. Be thankful for unforeseen gratuities, but also realize you may indeed deserve them. Avoid squandering unanticipated windfalls, but rather plough them back in to further enrich your future. Those in receipt of good deeds are most obliged to perform them.

Chapter Ten
TRANSITS TO LUNAR RETURN PLANETS

Moon Conjunct Lunar Return Sun

The heart of the matter and the main focus of the month gets its greatest boost now, even if it may not seem so at the moment, so keep your eyes wide open and be ready to jump to the tune when it starts to play. It's a time to whip up enthusiasm, but not necessarily to get down to brass tacks, so put on a happy face and spread good feelings around, even if nothing else seems to get done. By engendering warmth and excitement, you focus the flood in your direction and partake not only of your own energy, but the environmental electricity flowing down your wire.

Moon Conjunct Lunar Return Mercury

This is a time for clearing things up and attempting to be particularly lucid about your current set of plans. New insights can be worked into the picture and details laid out to those who need to be in the know. Bounce ideas off other heads to see a less subjective view of your schemata, then proceed to set things into motion before you lose the freshness of the inspiration. Though feelings may compel, you're best off right now with only what can be expressed clearly in words.

Moon Conjunct Lunar Return Venus

It's a good time to present things in the best possible light and dress up your reality in its Sunday best. When you look good, you feel good, and you make others around you feel better as well. By surrounding yourself with the trappings of love, desire, wealth, and beauty, you bring those very same elements closer to you, so indulge yourself in order to attract the future wherewithal of self-indulgence. In other words, consider play to be an investment in the future, as one good thing leads to another.

Moon Conjunct Lunar Return Mars

A quick, deft thrust can slay your quarry, but shooting from the hip risks overwhelming return fire. Rash actions seated in annoyance or anger can do a lot more damage than just sitting on your hands and waiting it out. Forceful action can carry the day, but only if it's well-thought-out and comes from a cool head, not a hot heart. The opportunity here lies in knowing when to pass and let someone else make the mistake, after which you may capitalize upon it.

Moon Conjunct Lunar Return Jupiter

Positive thinking and a generous imagination can now set the stage for increased opportunity and a bigger piece of the pie. The only limits you have are those you impose upon yourself, and now is not the time to do that. Think big, not only for yourself but for those involved with you, so that win-win situations can blossom and lead to ever-expanding possibilities for all concerned. When you invest in others, you invest in yourself, paying dividends to your posterity.

Moon Conjunct Lunar Return Saturn

Setting limits and downsizing some of your operations may be just what you need to get a tighter focus and better utilize limited resources. This may at first seem a downer, but it's really a matter of putting quality over quantity. You cannot be all things to all people, so you have to select what role you can effectively play and divert your assets to support it. Carefully chosen sacrifices allow all your guns to bear on the target you can't afford to miss.

Moon Conjunct Lunar Return Uranus

Taking risks on sudden, far-out ideas may generally be inadvisable, but this may be just the time to do it. Sudden flashes of insight may cut through the normal, humdrum solutions and start you thinking out of the box. Don't just throw caution to the wind, but don't hesitate to try something new and different this month if it seems appropriate to the occasion. Remember, however, that doing so may change more than you think and could leave you off balance unless you take a flexible stance.

Moon Conjunct Lunar Return Neptune

Trust your feelings and try to harness your intuition, which is favored. Although goals are different from dreams, you must dream to have goals, and it's at times like this that they are born. The seeds of the future are borne on the wind, and when reveries come to roost, the fantasies of tomorrow become the accomplishments of today. But like a hall of mirrors, only one of the images is you, and you will have to find out which one that is, in order to make it come true.

Moon Conjunct Lunar Return Pluto

You may have to briefly force someone's hand in order to win this round, but that's all part of the game, so don't flinch at the possible necessity. Also don't be surprised if turnabout is fair play and you find yourself the subject of a power play. The bottom line is to remember that it's more than just your game and you can only win if everybody gets to the finish line. Don't try to overwhelm pockets of resistance, however, as a Pyrrhic victory is an empiric defeat.

Moon Conjunct Lunar Return North Node

This is a time when the fox comes into view and the chase is begun, but you may not want to participate in the hunt, so take stock before you go galloping off. The opportunity for emotional or financial attachments and the responsibilities they entail may be something you want to pass on for the moment. Today's blithe commitment all too often becomes tomorrow's burden, so be sure that you really want it before you saddle yourself with more than you may want to carry. This does not mean you should avoid making commitments altogether, but that many call and only a few should be chosen.

Moon Conjunct Lunar Return South Node

This is a time when the whale has sounded and you don't know just where it will surface, including directly beneath you like Moby Dick. All kinds of unpredictable events, big and small, fortuitous or disastrous, seem to be around the bend, but chances are you'll only hear the waves lapping at the sides. Still, you're best off remaining in a state of preparedness for the moment so you won't be blindsided by either challenges or opportunities that surface unannounced.

Chapter Eleven

FULL AND NEW MOONS IN NATAL AND LUNAR RETURN HOUSES

FULL MOON IN NATAL HOUSES

Full Moon in Natal First House

For a day or so you may find that things get fairly wacky in your surroundings, and one of the causes may be you. It's a time when the time is right (and the pressure is on) to bring critical projects to a conclusion, put on finishing touches, and show the world just what you're all about. The emphasis is not so much on what you've done, but rather on who you are, and it's the right time to let your personality out of the box and shine, shine, shine. The whole set of circumstances may leave you a little giddy, but that's part of the surrounding picture at Full Moon, and this is your time to take a ride on that energy and put your personal mark on what comes of it.

Full Moon in Natal Second House

It's time to wrap things up financially, collect on projects that have come to fruition, and generally tie up loose ends. That should mean ending extra spending for a time, as it will be less necessary, but that's hard to do at a Full Moon, when things are a little crazy and judgment isn't always on the mark. Still, put your focus on that last payment, that final accessory, as the appropriate approach is to round out what you have done, put on the final touches, and then showcase it. It's a time to show off what you've got and make the most of it, not keep piling on to it. From now on, you should concentrate on reaping its rewards.

Full Moon in Natal Third House

Don't everybody talk at once! That's what you're likely to be saying as the volume level around you peaks and everybody is trying to get a message through, including yourself. Full Moons are like that in general, but this time it can be especially wordy as you try to finish up a lot of correspondence and tie up loose ends as the links you have established begin to really amount to something. That means crowding a lot of important last-minute things into the space of a few days, so don't overschedule yourself. Leave plenty of room open for situations that come rushing in and demand attention. Concentrate on matters that require conclusion and closure right now, and save your new ideas for later when they can get the proper attention.

Full Moon in Natal Fourth House

You may find that you could use a little more space at home as everyone there, including yourself, is taking up a lot more psychic space than usual right now. This can be claustrophobic, and you can get on each other's nerves—or you can use it as the opportunity to connect more by opening that personal space and letting people in. Or, you can just get out of the house and take a break from it all. Whatever you choose, you will find yourself with matters coming to a head, and you'll be making conclusions about just how well your internal game plan is working out. Literally and figuratively, good fences make good neighbors, but only when located strategically. It's time to finish yours, and then test it.

Full Moon in Natal Fifth House

Spontaneity is the watchword right now, and it's time to kick back and have some fun, especially with close friends and children. If you're at a party and go just a little bit over the top, you will likely be in good company, as this is a general time of celebration for all. Passion and intensity, on any topic you choose, runs rampant and enthusiasm overflows. If you're trying to get business done, you'll be swimming against the tide, so if you have the option, just go with the flow. If you're working on a creative project, make sure you get the product preserved for posterity—on paper, film, tape, etc. If you're just going to be pleasantly out to lunch, don't bother, just enjoy.

Full Moon in Natal Sixth House

If you aren't finally achieving something from your diet or exercise regimen, ask yourself why not. This is the time when you should be really getting things together and it should be starting to pay off. However, avoid excess and going over the top right now, as you may feel you can do more—and maybe eat more—than you really should. At work, you may feel overbooked, as tasks multiply and time is short to complete them. Once you've put on the finishing touches, however, you'll be able to stand back and take pride in your accomplishments. This is a good time for an office party, if you can concoct a good excuse for one. In general right now, celebrate finished tasks.

Full Moon in Natal Seventh House

This is a wonderful time to let your partner shine and just lay back and appreciate it all. That can mean enjoying the fruits of your mutual labors wrought by the relationship, or it can mean finally fully connecting with someone you've been developing something special with for a while. In either case, the rewards of connecting closely will take center stage, and you can lean on the familiarity you both already have to support what comes next. This can apply equally well in business as in romance, as long as you're dealing with an equal on equal ground. The essence is that you bring things to full bloom right now, so you can begin harvesting what you have sown.

Full Moon in Natal Eighth House

If you've been chasing down that extra credit line or looking for a better deal on a mortgage or equity loan, this is a likely time to pull it all together, as long as you've got all your ducks in a row. That may mean scrambling for some last-minute details or information to make it under a deadline, but there's real potential here. It's also good psychic weather for putting the finishing touches on restoration projects, picking up that antique you've been chasing down, and clearing out the cobwebs that may have been settling on lesser-used items. You'll have the courage to toss out those useless things you've been hoarding, and the burst of freedom will do you good. The theme: Finish clearing the decks for action.

Full Moon in Natal Ninth House

Think big, stretch the imagination, and push the envelope of your beliefs and feelings—this Full Moon has the potential to be an eye-opener. The energy is in the air to expand your window on the world, especially where it has been an ongoing operation already in the works. The heady feeling of suddenly having put it all together can have you wondering now what to do with it all, but that will take care of itself down the line. Look for good news on the legal front, as things wrap up and come to a conclusion, but don't count your chickens yet, as it's easy to overestimate right now. Main issues can get resolved now, as all parties are willing to go the extra mile to agree in principle, if not on the details.

Full Moon in Natal Tenth House

This may be the perfect time to nail down career matters that have been under development for a while. Although the pace may be accelerated and last-minute additions come crowding in, you can have a fuller, rosier picture of just where you stand in your profession and what people think of you. Choose this time to release a shotgun blast of self-promotion, whatever it is you do, as public-relations efforts have extra carrying power from this Full Moon. The buzz about you comes in a rising crescendo that peaks now, so start figuring out ways to take it to the bank. Although you may be tempted to squirrel yourself away at home, take time out these few days to wave your own flag.

Full Moon in Natal Eleventh House

Relaxation, comfort with your surroundings, and the warmth and glow of friendship can be the watchwords of this Full Moon, if you're willing to disconnect from the daily grind and tap into your personal happy side for a while. Fine food, pleasant company, tales of shared lives, and mutual adventures bring a fullness that makes you realize that the good life is about feelings, not money, friendship, or possessions. Satisfaction comes from just letting things happen for a moment and inviting in the blessings that still surround you, despite the rigors of existence. If you want a peak experience, that's the one that's in the wings right now, when you're ready to roll it out. Take the time to put the seal on relationships that elevate you and support you.

Full Moon in Natal Twelfth House

Although this Full Moon may at times seem somewhere between dream and hallucination, it emphasizes the ability to perceive and commune with your inner self in ways you might have been cut off from. The intensity in the air breaks internal barriers and allows you to heal wounds, now that they have come into the light. In more mundane affairs, things being cooked up behind the scenes come to fruition, although you may have to take extra precautions that they don't come into the light, unless you are fully ready to go public with them. Avoid confrontations, which will put you at a disadvantage right now. Use subtler means to gain your ends when emotions run high.

FULL MOON IN LUNAR RETURN HOUSES

Full Moon in Lunar Return First House

You can find things at their most advantageous now, when you can pull it all together by sheer force of personality. You're likely to have solid support from partners if you require it, and they will let you take the lead if you appear to be headed for a conclusion. Your apparent ability to wrap up situations begun a couple of weeks ago will be the indicator of your power, so you should sell yourself as a unifier who can bring disparate parts into a finished whole. Don't feel you have to succeed at it 100 percent, as your role may be quite different next month, but give it your best shot and see if you are comfortable filling out this persona. The catch phrase that fits: test of leadership.

Full Moon in Lunar Return Second House

Your opportunity is here to close on money arrangements of any sort. You are better off concentrating on short-term arrangements as you may want to change your mind later and write the situation off as a test run. So, wherever you can, if you make a final commitment to buy or borrow, make sure the terms run out soon but are renewable, or make sure you've got a guarantee and can return it if it doesn't work out. Chances are you can find cooperation in doing this, so don't feel you are asking too much. As the customer, you have the final clout and should not hesitate to exercise it. By doing so, you can also get a better deal on price and interest.

Full Moon in Lunar Return Third House

Your ability as a communicator may be put to the test right about now, and to the extent that you can gather and integrate disparate opinions you will succeed in taking charge of the situation. By promising a single, resolving approach to a variety of inputs, you can put yourself in the driver's seat, or at least be the one everybody comes to when they want a definitive opinion. Be a pollster and take the pulse of those around you, then help everyone get on board of whatever seems to be the prevalent and favored approach. You may have to let your answering machine take calls, as there can be an overflow and you'll want to have the strategic choice of who to call back first.

Full Moon in Lunar Return Fourth House

Finish up shorter projects underway on the home front and save larger ones until next month. What you launched two weeks ago should be wrapped up by now. Although you may have successfully siphoned off time from pursuing career matters, it's time to get on with it and move back onto a regular schedule. Don't hesitate to use a quick fix if necessary, as you may not be too sure of what you may have done in haste. Leave the option open to change your mind. You may be tempted to draw your bottom line over some seemingly important issue right now, but make sure it's a line in the sand and not in concrete. Things will be different next month, don't box yourself in.

Full Moon in Lunar Return Fifth House

You may be called upon to pick your pleasure right now, so it's time to focus on the person or pasttime that is really your month's fave. That means you don't have to give up all the rest forever, just put your full attention on one right now and see how doing it to the fullest works out. That can mean concentrating on one child, picking one sport to play, visiting one interesting place, indulging in one kind of food, exploring one style of music, and so on. It's time to pick and then try pursuing one all the way. If you are really creative, you can combine several at once so they work together, one reinforcing the next. In general, it's playtime, but follow through with it, don't just dabble.

Full Moon in Lunar Return Sixth House

Schemes and schedules you may have set in motion at work are likely to be tested and found either successful or consigned to the scrap heap. The former brings praise, and fortunately the latter will soon be forgotten. This is also a good time to take a close look at what you're eating and what kind of exercise you've been getting—maybe too much of one and not the best of the other, or maybe it's all coming together as planned. At any rate, this will be the time to declare it worthy or chuck it and try a new approach. This is a great time to make an assault upon and weed out destructive habits, recurring annoyances, and bad scheduling, and streamline your life.

Full Moon in Lunar Return Seventh House

If you've been playing the field, now is the appropriate time to settle down and concentrate on one person. If you're already attached, it's the perfect time to bring together the ongoing threads in your relationship and tie them together into one. Togetherness and fullness are the watchwords, and enjoy them while you may, as you may be focused on something entirely different next month. If you can't be with the one you love, love the one you're with, as it's the principle of the thing that counts right now, not achieving permanence. The same goes for the professional world, where you can consummate deals and cozy up to prospects without tying yourself down forever.

Full Moon in Lunar Return Eighth House

Wrapping up short-term credit arrangements is advisable, which can mean paying off or cancelling credit cards or loans, and finalizing arrangements in the works for accessing further resources. You will want agreed-on provisions to be as flexible as possible so you can have your cake and eat it, too. Leave loopholes on your side, but not on the other. This can also be final decision time as to what clutter you want out of your life, but don't give or throw things away when you can put them on extended loan to another and thus still have the option of getting them back. That way you can clear your decks of detritus, reap gratitude, and lose nothing.

Full Moon in Lunar Return Ninth House

It's time to declare earlier brainstorms and pipe dreams either feasible or not, and then put them into operation or junk them. Those ideas that have stood the test of time will remain, and the rest will pass. This can apply equally to legal obligations, politics, or personal beliefs. Either they work for you or they don't. That doesn't mean you should register in the opposing political party or convert to another faith, but you might try voting a different way and actually taking part with others in a different belief system. In other words, it's time to cleave to one position and see if it works for you, rather than scattering your opinion. Be opinionated, but not judgmental; be righteous, but not self-righteous. Feel what it's like to stick to your guns, and then you'll know if they are the guns to stick with.

Full Moon in Lunar Return Tenth House

Coming to some final conclusions about the current direction of your career would be in order now. This is not the time to make permanent decisions or try radically new directions, but analyzing just how successful your strategies have been would be a good idea. This especially applies to observing how others view what you are doing, as external cooperation and support are essential to shoring up even the greatest accomplishments. If your efforts have not been bringing applause, it might be time to rethink your strategies and pick a more crowd-pleasing approach. You bank on your reputation more than you might think, so make sure there is sufficient currency in your account.

Full Moon in Lunar Return Eleventh House

If you have embarked on opening new doors and currying favor with those especially situated where they can help you, make your move now to see if you have succeeded. Get them to come through with what they seem to have been promising, or get on with it elsewhere. If you score, so much the better—if not, there are lots of good fish in the sea. You've got time to find the right people, but not to waste with the wrong ones. Do not, however, burn bridges, as you could need them later. In any case, remember who your real, close friends are and take the time to re-seal your affection and loyalty with the ones you know you can count on.

Full Moon in Lunar Return Twelfth House

Skullduggery at work may be hatched behind the scenes, and this is certainly a time for bringing undercover operations to fruition. Keep your eyes open so you can avoid involvement, though backdoor operations are not always ill-construed and some kinds of good are better done in the dark. In a more personal vein, pressures and problems that have been building up may burst out into the open, so take the opportunity to field them and defuse them before they again go underground. Out of sight is not necessarily out of mind, and gnawing issues don't often have this opportunity of being addressed, so be open about yourself and especially tolerant of the upsets of others.

NEW MOON IN NATAL HOUSES

New Moon in Natal First House

This New Moon marks a time of personal beginnings, when you find yourself shot into the limelight and all eyes turn toward you to see what you're going to do next. This may be somewhat startling, as it's rather different than what last month brought, and you will need to step up and display your self-confidence in a way that wasn't called on before. In a volatile situation, you will need to have your wits about you as you will have a particularly high profile. The trick is to handle it with grace and not let it go to your head, as this, too, shall pass, and you'll want to look back on a level performance that maximized the opportunities dropped into your lap.

New Moon in Natal Second House

For a day or so you may be under particular pressure to underwrite new projects or financial situations that require sinking money into. It may seem quite urgent, but it's a time when everything seems a bit more intense, so you may do well to stand back and take a second look before jumping in with both feet. Wise investments made at this time can bring good returns in the next couple of weeks, but ill-conceived spending could drain you in the same time period. Know you're right, then go ahead one way or the other, as the one option you probably won't have is sitting on the fence and putting it all off until later. Strike when the iron is hot or pass the opportunity on to another.

New Moon in Natal Third House

A flurry of activity for several days can have the phone ringing off the hook as new ideas and propositions come your way that require either your opinion or your decision. Your challenge will be sorting the wheat from the chaff on the fly, and sorting out the confusion that always abounds at the New Moon, especially this one. The day is won by multiple, smaller choices made wisely, rather than hanging on a single issue, so don't obsess but decide quickly and move right along to the next. This is a great time for making new acquaintances as the air of intensity makes for strong personal connections and breaks through the barriers of unfamiliarity.

New Moon in Natal Fourth House

This may mark a day or so of raised tensions at home as biological tides run high and elbow room seems harder to come by. Used creatively, however, that can lead to new ways of organizing and arranging your personal space so it doesn't overlap that of others. It's a good time to start new home-related projects as a way to utilize the extra energy in a constructive manner. Letting out your deeper expressions of feeling to another can make things especially intimate right now, but make sure you express yourself clearly and are not misunderstood. This is the perfect time to feel "at home" with someone, either literally or figuratively. The only thing you won't want to be is idle—the vibes are too intense to just sit on your hands.

New Moon in Natal Fifth House

Find a party to go to, give one, or just party down where you stand—the astral weather is perfect for it. In fact, anything you do for enjoyment will seem twice as good right about now. Play with your children, or just break out the inner child, as you can more easily part with self-consciousness and spontaneity breaks out, well, spontaneously! If you call on the Muse for a creative project, she will definitely show up (Muses love parties and children), and you'll find that your juices flow faster and easier than usual. There can be a fine line between joyous adventure and risky behavior, however, so try to stay on the safe side of the line. This is a good time to engage, but not be, a designated driver, both literally and figuratively, so you can be safely irresponsible for a while.

New Moon in Natal Sixth House

This is a good time to take your blood pressure—if it's not elevated now, you're in really good shape. It can be easy to worry about health matters now, but moderation is the right response. The temptation to throw yourself into a new regimen may be great, but choose your new path well before you embark on it or it will be a flash in the pan. The same goes for cleaning house in the workplace and starting off on a fresh footing. Make sure you're not demanding too much of yourself in the long run—you will not always have this much drive, and you don't want to peter out because you can't keep up with yourself. A good program is determined by whether you can maintain it on an off day.

New Moon in Natal Seventh House

This is a particularly ripe couple of days for finding a new partner or turning over a new leaf with an old one. You may find a flood of emotional intensity coming your way, and how you utilize it could largely shape the direction of the relationship. That means be particularly understanding and realize that though feelings may be overstated at this time and emotions run hot, the message underneath is sincere and needs to be taken seriously. Breakthroughs can occur now that you both might have kept a lid on otherwise, and you'll be glad you achieved the new closeness, even though you may not be able to keep it going at that level of intensity all month.

New Moon in Natal Eighth House

You may find that turning over a new leaf this month entails tossing out the old and ringing in the new, but you should take a second look before doing that. You may find it much more satisfactory to simply rework and refurbish what is already there. Sometimes a new paint job makes everything new again, so you don't have to totally start over from scratch. They don't make things like they used to, so preserve what you can. It's also a good time to initiate that new credit card, but avoid impulse buying, as it's easy to go overboard right about now. Ditto for mortgage or other loans—this is a good starting time, but just take what you need and no more.

New Moon in Natal Ninth House

It's a great time for planning, cooking up new schemes, and rearranging the way you look at things. Patterns you establish now do not need detailed follow-up until later, so you can stick to generalities until you've got it all properly roughed-out. Put on finishing touches later. New publishing efforts, or any project that generally disseminates what you have to say, are well begun now, and you may find that the pressure is on to produce. Avoid hasty commitments, however, as you don't want to be held to something ill-considered or not fully thought out. Well begun is half done, so be thoughtful, deliberate, and measured in your efforts, even though you may be tempted or even encouraged not to be.

New Moon in Natal Tenth House

New developments in career matters may have you quite wound up for a couple of days, and the pressure to produce can be intense. Too many things happening at once makes it hard to keep track of things, but you can't afford to leave anything out. It is as important to listen at this time as to act, because what people are saying to you and about you very much reflects your personal and professional status, and you will want to make sure that you are being represented honestly. If you give your resume a second look, you'll find there are new things to say about yourself now that need to be included. Taking a fresh stance can put you into new places and move you up a notch.

New Moon in Natal Eleventh House

Feelings run high among close friends, and an intensity is available right now that can renew the love and passion that first fueled your relationships. It can be worth risking upset to get at what you are really feeling and share it with those who matter the most. Because there's a crazy atmosphere surrounding it all, you can say things you might have hesitated to bring up and wear your heart on your sleeve without worrying about the outcome. The same ability to burst out of your shell will also aid you in getting into new company whom you might have thought didn't care about you before. An approach of disarming honesty can make you welcome in unexpected places and bring you valuable allies.

New Moon in Natal Twelfth House

Eruptions of the unconscious could bring both troubles and personal revelations to the surface right now, so be careful not to betray secrets or let the cat out of the bag prematurely. Don't say the first thing that comes to mind, or you may give away more than you intend to. New deals are being struck behind the scenes with a strong sense of urgency, but they are not necessarily justified, so watch that you don't climb aboard the wrong bandwagon right now. Wait until things come out into the open to make decisions or take sides, even though there appears to be advantage in secrecy. Open covenants, openly arrived at, are the better part of honor.

NEW MOON IN LUNAR RETURN HOUSES

New Moon in Lunar Return First House

A brief period when this month's personal issues come into focus is underway, especially where you can be tagged as an initiator. You have the option of establishing your personal picture for the period by asserting yourself at this time, and you can expect involvements you throw yourself into right now to come to at least a temporary head in about two weeks. You'll be best off keeping your head about you when others are losing theirs, but don't feel you can't get out of what you've done now, as it will be a new ball game next month. Nevertheless, if you want the spotlight, it can be yours for the moment, but be ready to pass the baton when it's time.

New Moon in Lunar Return Second House

These few days are favorable for making down payments, signing new contracts, and taking new approaches to earning, especially those that require only short-term involvement. Expect initial payoffs about two weeks later from financial commitments you make now. Try making this the time you nail down your budget for the coming month and then make a habit of it around this time each month. This may also be the peak spending period for the month, but that depends how you budget things. Don't feel you are nailed to what you get into now, but see how it works out in the coming weeks and then decide if you want to continue in this pattern or if you want to take a new tack on where and how you spend your money.

New Moon in Lunar Return Third House

Talk is cheap right about now, but also holds promise. This is a good time to test out new ideas on people and make new contacts and acquaintances. Trial balloons launched now that fizzle later will be soon forgotten, but proposals that fly will be the subject of redesign and finalization later, so you can feel free to make mistakes, experiment, and take mental excursions without the risk of having them drag you down later. By the same token, don't expect new people with whom you engage now to turn into long-term friends or resources, though they will help expand your reach and your Rolodex, so count it as a plus. Your real eggs may be in another basket, but it's good to see what's around.

New Moon in Lunar Return Fourth House

If you aren't already working from home, you might have the opportunity to try it out for a few days and see how it suits you, even if it only means taking work home with you for a while. In general, experimenting with new kinds of personal arrangements and lifestyles at home can put some spice into the old routine and refresh your inner landscape. This doesn't (or shouldn't) require making permanent changes, but rather initiating a new approach and seeing how it fits you. In other words, rearrange things, but make sure you can put them back the way they were until you've decided just what pattern you really like, giving special attention to how secure it makes you feel.

New Moon in Lunar Return Fifth House

Games can be played for the fun of the moment or they can be played seriously, and right now is a good time for the first. Try out new pleasures and new enjoyments, whether physical, mental, or spiritual—anything that fuels your creativity and introduces you to something new and fun. Go on an adventure with your kids, take up a new sport, get artsy-craftsy, or pick up an instrument and play it. Add some variety to your love life, become a connoisseur, or be the life of the party. The opportunity to try new things to see if you like them is knocking, so opt for the free trial, no obligation required. But, of course, if you like it, you can subscribe.

New Moon in Lunar Return Sixth House

If you always wondered if you should try that off-the-wall diet or take that exercise trip, now's the perfect time to try on the shoe and see if it fits. Give yourself two weeks to see how it goes, and then decide if you want to continue. You never know, and right now you don't have to commit, so why not? The same goes for trying out new approaches at work. If changing the schedule seems appropriate, run it up the flagpole and see who salutes. If no one does, no one will particularly blame you, but if it turns things around, you can take all the credit for the idea. Dabble in the new without having to be an expert, and see what happens.

New Moon in Lunar Return Seventh House

If ever there was a time for a blind date, this is it. The opportunity to try on someone new can present itself, and you can do it without commitment, without obligation. That doesn't mean irresponsibility—quite the contrary; if you've got nothing to lose, you're much more likely to relax and enjoy the freedom to be yourself. That way, you'll get an equally honest response. Even in established relationships, you'll find you can both try on new hats without having your roles reassigned, and that can go a long way toward better mutual understanding. If you don't like it, don't repeat it—but if you do, you'll have something new on your personal palette. The operative description: Take a ride; it's free.

New Moon in Lunar Return Eighth House

Everything old is new again—or, throw out the old and ring in the new. Chances are you can try out both right now without having to spend too much on the new or permanently get rid of things you really might need. So, it's a good time to dabble, try your hand at modernization, or pull something out of the closet and see if it fits again. Play with new approaches, old chestnuts. You can do the same with temporary financial obligations as well, and don't be shy to show up at the returns counter in a couple of weeks to get a refund. On the other hand, you could luck into just what you need—then you'll be glad to pay for it. Either way, no regrets.

New Moon in Lunar Return Ninth House

Getting into blue-sky brainstorming and giving even the most crackpot ideas a run for their money can provide fun and maybe even profit now. Original thinking is the better for being temporarily free of ties and responsibilities, so scheme on, however unusual it may seem. Spend the next couple of weeks following through on what you've dreamed up, then decide if it's garbage in, garbage out, or the makings of the Nobel Prize. Explore some of the areas where they give out Nobels: peace, literature, economics, or science. All of these can use some fresh blood and inspiration—why shouldn't it be yours? If it turns out to be really wacko, toss it. If not, patent it.

New Moon in Lunar Return Tenth House

What would you really like people to say about you? If that's not what they're talking about, you could try a paradigm shift in what you do for a living. Don't quit your job and move to another country, just make some moves to head in a whole new direction. Then tell people about it and see what reactions you get. They may think you're crazy, but they may also say it's great, they always thought you had that in you. It might even convince you to run with it, or it might not. But what it will do is give you a fresh look at yourself as others see you, and that is a large part of what determines your professional success. Maybe you've been missing your calling (or just the right approach to it) all along, and all you had to do was ask for feedback.

New Moon in Lunar Return Eleventh House

Take a new tack with your close friends this week and you may find that they are resources for you in ways you had not imagined. You can't find out if you don't ask. Propose new ways of getting together and examining and developing your relationships that might add miles to what you do. You will find out a lot without risking a lot—the worst they can do is say no. Similarly, don't be afraid of making a fool out of yourself in front of your superiors—it's not going to happen, and it could show you off in a new light that will bring extra support from unexpected quarters. Wearing your heart on your sleeve can get you everything, with little chance of harm, so don't hesitate to do so.

New Moon in Lunar Return Twelfth House

Make life your very own inkblot test. Try free associating and just saying what comes into your mind. How do you really feel about something, someone, a situation, yourself? Don't cogitate, regurgitate. First impressions are often the deepest, so try going with them for a while. If it devolves into confusion or depression, drop it—but if it reveals sudden insight, you're a winner. Similarly, a little spontaneous plotting behind the scenes right now can't hurt too much, as you can abandon it or even disavow it later. But if your plans work out, you could be the power behind the throne. What's buried can begin to tunnel out right now, surfacing in about two weeks.

Chapter Twelve
TRANSITING RETROGRADES AND ECLIPSES

Mercury Retrograde

Three times each year Mercury goes retrograde for three weeks, impacting one or two of your Lunar Returns. The period during which it is retrograde marks a background that affects everyone in a similar way and is not to be ignored. It's a period of some confusion when messages and meanings go awry, and even with the best-laid plans you discover that something was left out or needs correction. Therefore, it is a great time for finishing things, putting on the last touches, or editing the final project, as you'll find mistakes you might have otherwise missed. Conversely, it is less than ideal for beginning anything having to do with communications, as you may build in errors without meaning to and have to spend time reworking it after it is launched.

Venus Retrograde

Every couple of years Venus goes retrograde for a little over a month, impacting one or two of your Lunar Returns. The period during which it is retrograde marks a background that affects everyone in a similar way and is not to be ignored. Projects involving earning or investing money will be slowed down or reversed, as will personal efforts toward getting what you need to make your life satisfying. Your best bet is not to fight the tide, but instead allow it to take its toll on the rest of the world while you hang back and wait for a better moment. If you're staying still and everyone else is moving backward, you're moving ahead. Rethink investments and beautification projects, including clothes and cosmetology.

Mars Retrograde

Every couple of years Mars goes retrograde for about two months, impacting two or three of your Lunar Returns. The period during which it is retrograde marks a background that affects everyone in a similar way and is not to be ignored. It slows down ongoing projects, often hamstringing the ability to get on with things. It forces restarts after the period is over, and it generally causes the energy level all around to decrease, dragged down by unnecessary repetition, frustration, and fatigue. This, too, will pass, but wise up accordingly, store up extra energy, and don't try to overshoot the mark. Wait until it's over to begin projects that you want to be especially strong and vigorous.

Lunar Eclipse

When a Full Moon is also a Lunar Eclipse, its impact is twice as important. Expect sudden turnarounds, unexpected new directions, and serious paradigm shifts in the house in which it falls along with the planets the Moon touches that day. The effect will have been building for the last month and will have effects running into the next five months. Because of its intensity, this particular part of your Lunar Return may quite literally eclipse much of the rest of the month's goings-on. Its effects fall mainly on the emotional side and ripple outward from there. Expect to make room for resultant happenings for some time to come.

Solar Eclipse

When a New Moon is also a Solar Eclipse, its impact is twice as important. The house it falls in and the planets it touches can expect a reversal of field that has been building for a month and will impact the next five months. The energies in these areas are subjected to tornado-like twisting that usually catches everyone by surprise, though you will see how obvious it was in hindsight. Because both the Sun and Moon are in the same spot, it has a very dominating and physical effect as well as an emotional effect, so you can expect some of the changes here to overshadow the rest of what goes on during the month.

ABOUT THE AUTHOR

John Townley is a pioneer of innovative and accessible techniques in professional astrology. He is the recognized "father" of the composite chart, astrology's most widely used modern relationship tool, which he introduced to America in the early 1970s. He is a former president of the Astrologers' Guild of America, was the editor of *The Astrological Review,* and is the author of many books and articles on astrological cycles, love, and relationships. A millennium Renaissance man, he is also an artist/producer of over a dozen record albums, the author of multiple scholarly maritime historical papers and book reviews, and an Internet multimedia journalist.

CPSIA information can be obtained at www.ICGtesting.com
Printed in the USA
BVOW051757160212

283130BV00003B/22/P

9 780738 703028